Arnold Lawson, a retired teacher, was head of biology and head of the science faculty in a large comprehensive school in Sheffield. From an early age and living on the edge of the English Lake District with its outstanding landscapes and diversity of flora and fauna, he has always had an interest in natural history. This interest has taken him from the forests of New Zealand to the Amazon jungle of Brazil. His other interests include gardening, fellwalking and landscape photography.

For Madeline, Kathryn and Hazel.

Arnold Lawson

MORE HORRIBLE BIOLOGY

AUSTIN MACAULEY PUBLISHERS®

LONDON ∗ CAMBRIDGE ∗ NEW YORK ∗ SHARJAH

A CIP catalogue record for this title is available from the British Library.

ISBN 9781035862092 (Paperback)
ISBN 9781035862108 (ePub e-book)

www.austinmacauley.com

First Published 2024
Austin Macauley Publishers Ltd®
1 Canada Square
Canary Wharf
London
E14 5AA

The author wishes to acknowledge with grateful thanks the encouragement, kindness and assistance of the editorial team of Austin Macauley Publishers. He also wishes to thank the kindness of his wife Madeline and his daughter Kathryn in encouraging him to write more about the unusual facts of biology and Hazel, his younger daughter, for sorting out his computer problems.

Table of Contents

1

Tongue Eating Louse

The tongue eating louse, which looks like a wood louse that you find hidden under stones and plant pots in gardens, is a harmless but horrible animal! It lives in the sea and is common in the waters off the coast of North America and California and the Atlantic Ocean. The male louse is only about 1.5 cm in length and the female about 2.5 cm, but it has a most unusual life. The male louse enters the fish by clinging onto the gills of the fish and then it changes into a female louse!! The female now mates with a male which is still on the gills. The female louse then crawls from the gills and into the mouth of the fish. The louse has very sharp claws on the end of its legs, and it uses these to attach itself to the tongue of the fish. The claws of the front legs of the louse now slice open the blood vessels of the tongue. The female louse now takes its first meal of blood! The louse also feeds on mucus and food particles in the fish's mouth. The female louse lays its eggs inside the mouth of the fish and eventually these hatch into small male lice. These remain inside the mouth and when another suitable fish swims past, the young lice shoot out of the mouth and become attached to the gills of the fish.

Because the louse continues to feed on the blood of the tongue, the tongue is eventually destroyed, and the louse must hold on to the root of the tongue. The louse becomes the tongue of the fish! What is unusual is that the fish does not seem to suffer even though it has lost its tongue.

The fishes which are invaded by these lice include snappers, grunts and croaker. Grunt fish are so called because they make a grunting sound when they grind their teeth together. Croak fish vibrate muscles around their swim bladder (an organ in the fish which contains oxygen and allows the fish to maintain its depth in the water) and make a croaking sound. In 2005 a snapper fish was found in a London's fishmonger's shop, and this may indicate that the tongue eating louse may be spreading across the Atlantic Ocean to the United Kingdom.

The tongue eating louse on the head of a fish. Note the sharp claws at the front of the animal for slicing open the blood vessels of the fish's tongue

The tongue eating louse in position of the tongue. The fish uses the louse as its new tongue.

2

Worlds Most Devastating Pest – the Locust

There are about eight billion people on our planet, but a large swarm of locusts can contain more of these insects than there are people on earth! One of the largest swarms of locusts ever recorded was in Africa in 1954. It was calculated to contain 10 billion locusts and in one day they ate 136,000 tons of vegetation including agricultural crops.

The desert locust is only about 3–4 inches long and lives about 3–5 months, but it is a ravenous eater of plants. After mating the female locusts lays its eggs in damp sandy soil. When the locusts hatch from the eggs they can form large swarms which can travel long distances in search of vegetation. The locust swarms occur in many parts of East Africa, the Middle East and parts of Asia including India and Pakistan. In 1988 a large swarm of locusts flew from Africa to the Caribbean, a distance of 3,100 miles (5,00 km) in 10 days. In 1954 a swarm travelled from Africa to the United Kingdom.

The worst problem with locust swarms is that they destroy agricultural crops as well as other vegetation, and this causes food shortages and even famine. Imagine a large swarm of locusts – this can be 20 miles long and 20 miles wide and hundreds of feet thick and contain 50 billion locusts. This swarm can eat in one day the equivalent of what 3.5 million people will eat in one day!! Locust swarms (sometimes called locust plagues) have been documented for thousands of years, but it was only in 1860 that the swarms were accurately recorded and since this date locust plagues have occurred every year from this date. Probably the greatest locust plague ever to take place was in Western USA in 1875 when the Rocky Mountain Locust reached unbelievable numbers. A doctor called Albert Child estimated the size of the locust swarm which eventually was called Albert's swarm. The locust swarm covered an area of 198,000 square miles (510,000 sq. km) which is four times the size of England! The weight of the locusts was estimated to be 27 million tons and contained 12 trillion locusts (a trillion is one thousand billion or 1,000,000,000,000). Strangely, the Rocky Mountain Locust is now extinct. After Albert's swarm, farmers planted their crops in the land where the locust laid their eggs and this destroyed the eggs of the Rocky Mountain Locust and the insect became extinct.

There have been several methods used to try and destroy these swarms of locusts. If the swarm is very small the natural predators of the locusts such as birds and lizards can easily eat them, or the farmers can remove them by waving clothing and cloths which dislodge them from the vegetation. However, for large and very large swarms it is far more difficult. Insecticides (chemicals that kill insects) are sprayed at ground level or from aircraft, but the insecticides are harmful to humans and the chemicals used are more effective against young locusts. One of the latest methods in controlling this pest is using a fungus, known as 'Green Muscle' which is sprayed on the swarm and eventually (after 7–14 days) kills the locusts. One advantage of this method of control is that it does not kill other insects or their predators.

A swarm of desert locusts.

The reddish-brown colour in the photo is part of a huge swarm of locusts.

Locusts resting on the branches of a tree.

A farmer try to protect his crop from the desert locust.

Desert locust infected with the 'Green Muscle' fungus as a method of trying to control this devastating pest.

3

The Stinkiest Flower in the World – Rafflesia or the Corpse Plant

Can you imagine a plant that has no leaves, no stem, and no roots, but has the largest and stinkiest flower in the world. The plant is called Rafflesia which lives in Sumatra and Borneo There are 28 different kinds (species) of Rafflesia of this very strange plant. The Rafflesia flower lives as a parasite on a plant called the chestnut vine. (Vines have very long stems that can either grow up trees or along the ground. The largest vine ever recorded was from Sumatra and was 1 km in length!!) The rafflesia flower starts as a small bud on the stem of the vine. During a period of about 27 months this bud will grow and eventually break open into the world's largest single flower. The largest Rafflesia flower recorded was in 2019 and measured 120 cm (4feet) in diameter and weighed over 11 kg (24 pounds). The flower obtains water and nutrients from the vine. The flower is either a male flower or a female flower. To produce seeds the flowers have to attract insects which carry the pollen from the male flower to the female flower so that pollination can take place. The flowers produce chemicals that have a most disgusting and obnoxious smell which is like rotten meat or that of a decaying corpse and the flower is described as the stinkiest flower in the world. This disgusting smell attracts bluebottles (a common fly found in homes and gardens in this country). The large Rafflesia flower only lasts for a few days so there is very little time for pollination to take place. Apart from having a horrible smell the flower has another trick up its sleeve, the flower increases in temperature that is higher than the surrounding air. This means that the vile smelling chemicals produced are distributed faster in the air and increases the chance of pollination. After the pollination process the female Rafflesia flower produces thousands of very small seeds. The bluebottles lay their eggs inside the female flower and these hatch into maggots. Unfortunately, the flower dies after a few days and there is no food for the maggots, and they die! The only way by which another flower can be produced is for a tiny seed to enter the stem of the grape vine. Biologists do not know how this happens. The Rafflesia is becoming extremely rare because the tropical forests where they live are being destroyed and the land used for farming. The buds are also collected for medicinal use. Biologists still do not know how to grow this magnificent flower from seed.

The Rafflesia was discovered in 1797 by a French explorer called Deschamp. When he was sailing back from Indonesia to France his ship was hijacked by a British ship and was at war with France at the time and his research notes and samples of the flowers were confiscated by the British. In 1818 a British surgeon called Joseph Arnold collected a specimen of the flower in an expedition led by Stamford

Raffles. The flower was then given the scientific name Rafflesia arnoldii.

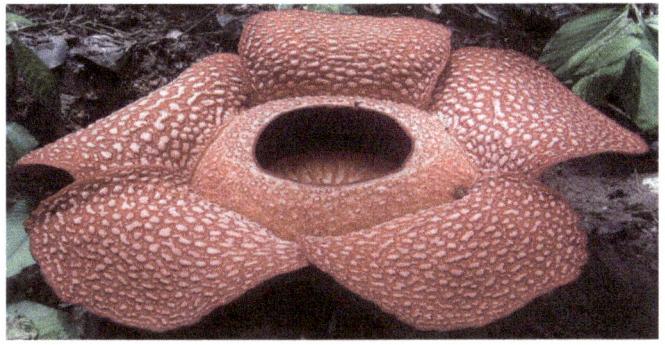

The Rafflesia flower which can grow up to 120 cm across. The plant has no leaves, no roots, and no stem. The flower grows as a parasite on the chestnut vine.

Three large buds of the Rafflesia plant can be seen in the right of the photograph.

4

Guinea Worm – Horrible Parasite, but Almost Eradicated

In 1986 there were over 3,500,000 people from over 20 different countries who were infected with a parasite known as the guinea worm. Less than 40 years later, in 2020, only 27 people from six countries were known to have this parasitic worm. It was at one time a common parasite in many parts of Africa, Asia and South America, but now only exists in isolated villages in the African countries of Chad, Ethiopia, Mali, South Sudan and Angola. Although there has been great progress made in the eradication in the world of really horrible infectious diseases, only one has been successful – smallpox, when the world was declared free of this disease in 1980.

People become infected with the guinea worm by drinking dirty stagnant water that contains microscopic animals called water fleas. Once inside the stomach the water fleas die, but inside the fleas are the larvae of the guinea worm. The larvae now turn into guinea worms, and they penetrate the walls of the stomach. Outside the stomach the worms mate. The female worm is about 80 cm (31 inches) in length, and the male worm about 4 cm in (1½ inches) and after mating the male worm dies. After about a year the female worm is found underneath the skin of the feet and legs and a very painful blister indicates the position of the guinea worm. The worm now leaves the body through the blister. As the worm leaves the body, the infected person often wraps the worm around a small stick and gradually pulls the worm through the blister, a process taking several days.

If the person with the worm leaving the skin now enters a place where drinking water is obtained, and the worm becomes wet it releases a large number of larvae into the water. If the water contains water fleas, they eat the larvae, and the infection process starts all over again.

There are no drugs or vaccines to protect people from this parasite. In 1980 it was decided that there should be an international programme to eradicate this horrible parasite. Within 40 years this programme had been highly successful, and this was due to very simple but effective rules for the infected areas. These included (1) only drink clean water, (2) filter the water to remove the water fleas, (3) cook fish thoroughly because they eat the water fleas, (4) do not feed dogs with uncooked fish because dogs also suffer from the guinea worms, and make sure that infected dogs are always tied up and do not allow them to visit clean water sources, (5) reward people who report infected dogs, and, (6) each village has a volunteer who is responsible for nursing infected persons and keeping records and passing on this information to health authorities. The infection caused by the guinea worm is close to becoming the world's second disease to be eradicated.

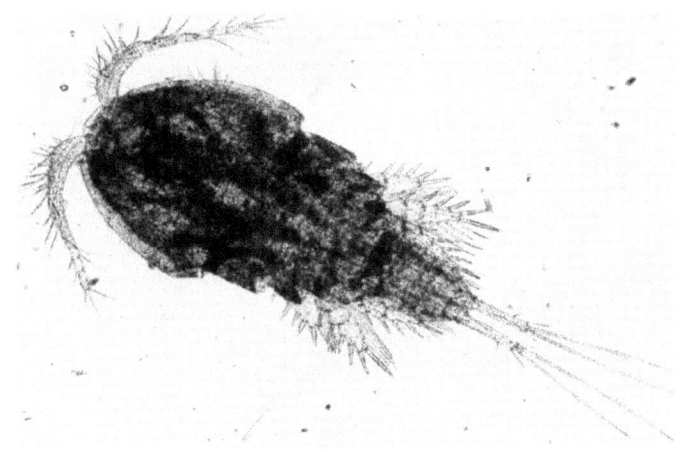

This small animal, called Cyclops only 2–3mm in length lives in water and feeds on the larvae of the guinea worm. If a person drinks water containing infected Cyclops, then the guinea worm passes into the person.

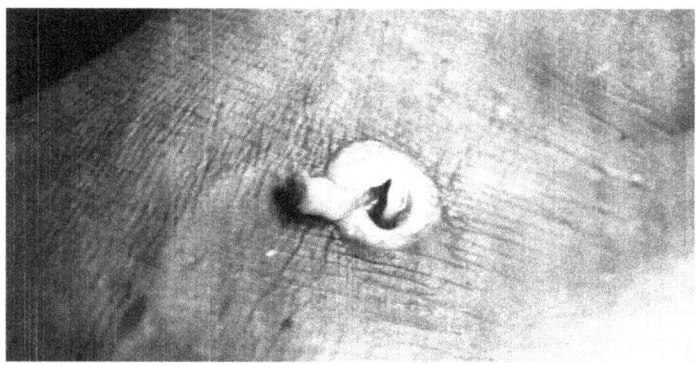

The guinea worm takes about a year to develop inside the infected person. A very painful blister appears, usually on the leg, and the worm begins to leave the body.

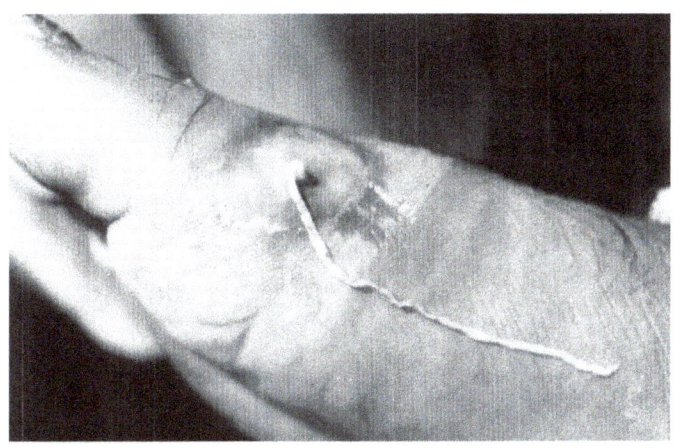

The guinea worm, about 60–100cm leaving the leg of the infected person.

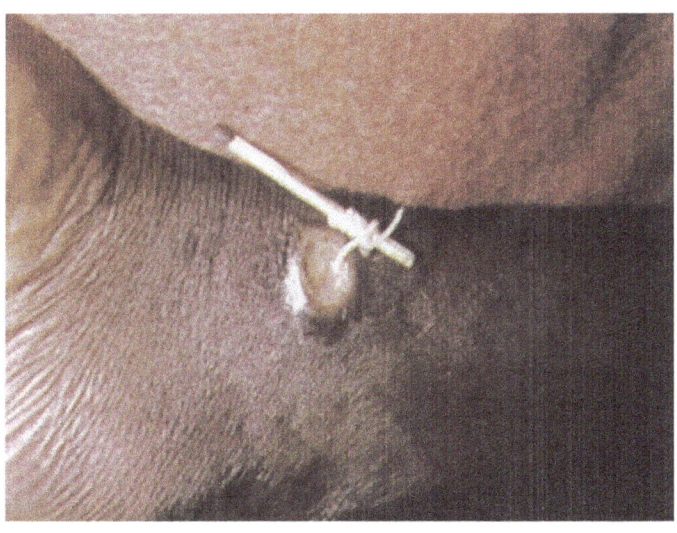

The traditional method to remove the emerging worm is to wrap the animal around a twig and carefully pull the worm out of the body, a process that can take several days.

5

Carnivorous Centipedes

Centipedes are fast, carnivorous, and venomous predators. There are over 3,000 species and 57 of these are found in the United Kingdom, where the largest is about 3cm in length. However, in some parts of the world some species of centipede can reach a gigantic length of 30cm (12 inches). The longest is the Amazonian Giant Centipede (sometimes called the Peruvian Giant Yellow Leg Centipede). Although the smallest centipedes feed on small insects, spiders and millipedes, the largest will capture and eat small lizards, small birds, mice and even bats. Centipedes are nocturnal creatures and during the day they live hidden underneath small stones, rotten wood, twigs and under decaying leaves. However, at night they are extremely active and will hunt down prey much larger than themselves.

The body of the centipede is divided into segments, each having one pair of legs. On the first segment of the centipede body the legs have become very sharp pincers called forciples which are used to capture and inject venom to kill the prey before it is eaten. Centipedes can be very aggressive and will bite humans if they feel threatened. Although only one person is known to have died from centipede venom, over

400 people a year attend hospital in Hawaii because of being bitten by the Golden Headed Centipede. The bite from all centipedes can be painful causing reddish swellings, and in general the larger the centipede the more painful the bite.

The Amazonian Giant Centipede sometimes lives in caves that are inhabited by bats. The centipede attaches itself to the roof of the cave by its strong legs with the rest of its body dangling in the air where it waits for small bats to fly past. If the bat flies to close to the centipede, then it is captured, injected with venom, and then eaten.

This man is holding the world's longest centipede at 30cm in length – the Amazonian Giant Centipede.

Centipedes are fearsome predators. In this case a centipede is killing and then feeding on a bird eating spider.

The Golden Headed centipede is very common in Hawaii. Its bite is extremely painful and causes a burning sensation around the bitten area. Although death from the bite is extremely rare, over 400 people a year attend hospitals in Hawaii for medical treatment.

A remarkable photograph of an Amazonian Giant centipede capturing a bat in a cave in Venezuela, South America. The centipede uses its back legs to anchor itself to the roof of the cave. When the bat flies past the front end of the centipede it is grabbed and injected with venom and eaten.

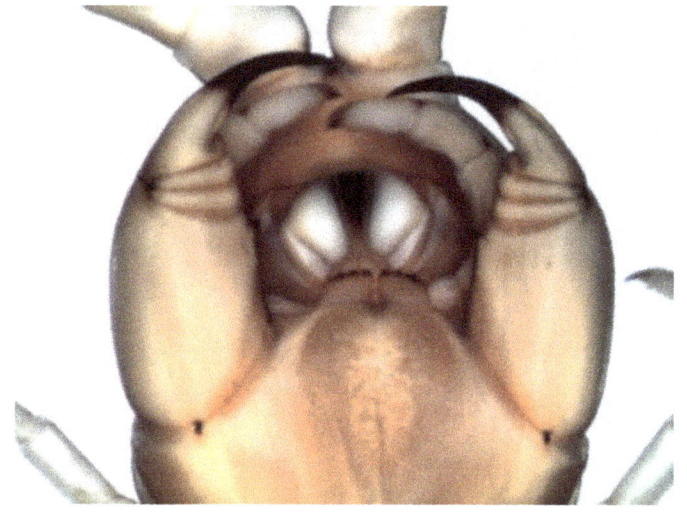

In centipedes the first pair of legs have become very sharp pincers called forcipules which are used to inject venom into prey.

6

Blood Squirting Lizards and Harvester Ants

There are over 6,500 species of lizards ranging in size from the 3m long Komodo Dragon found on Komodo Dragon Island, Indonesia to the smallest lizard the Nino Chameleon which is only 1cm–2cm in length and found in Madagascar.

Animals have ways of defending themselves against predators. The commonest form of defence is camouflage – animals blend themselves into the background and are very difficult to see. Another method is to remain stationary. One of the most bizarre methods of defence is to be found in the Greater Short Horned lizard.

There are 21 species of horned lizards and eight of these can squirt blood. Sometimes the blood is squirted towards the mouth of its predator, but the lizard often waits until it is in the mouth of the predator before ejecting its blood. The blood has the foulest taste and is very effective against cats, dogs, and coyotes that prey upon this lizard. Strangely this foul-tasting blood has no effect against some birds of prey which eats this species, e.g. burrowing owls.

Many of the horned lizards feed on harvester ants, so called because they collect or harvest seeds and carry them to their nests which are underground and where the seeds are eaten. About 1.5cm in length they are formidable insects and have the most dangerous venom of any known insect. The venom is more than 35 times more potent than that of the Western Diamondback Rattlesnake. Although the Horned lizard feeds on the Harvester ants the venom has no effect on the lizard. This venom is found in the blood of the lizard, and it is this venom that gives the blood its foul taste and helps the lizard to defend itself.

The Greater Short Horned lizard is about 7–9cm in length and lives in North America. It is well camouflaged, but it still requires another method of defence against predators.

The lizard feeds on harvester ants. It is a "sit and feed" predator waiting outside the nest of the ant and consumes about 20–60 insects per day. The ant has formidable mandibles which it uses to crack open the seeds upon which it feeds. The photograph above shows the stinger penetrating the skin of a person and the structures of the stinger being pulled away. The person will have severe pain for about 4 hours.

The greater Short horned lizard can accurately eject blood from its eyes towards a predator up to a distance of 1.5m. The foul-smelling blood contains the toxic venom from the harvester ants upon which it feeds. Sometimes the lizard does not release the blood until it is in the mouth of the predator! Because the blood has a foul taste the predator releases the lizard from its mouth.

7

Bloodletting and the Death of George Washington

On the 11th December 1799 the American president, George Washington, went horse riding for many hours on a cold wet miserable day with rain and snow. When he returned, he did not change into dry clothes because he did not want to be late for dinner. Two days later the 13th he took to his bed with a cold, a sore throat and difficulty in breathing and his wife Martha sent for his doctor. As was very common at the time George Washington was bled.

Bleeding was carried out in two ways. Firstly, medicinal leeches which feed on the blood of animals would be placed on the skin and the leeches would draw blood from the body. Secondly, a doctor (qualified or unqualified) would cut open the skin and a blood vessel and drain blood from the body. It was believed at the time that bloodletting would remove dangerous substances out of the body and so reduce the illness of the patient. There is much controversy about the number of times George Washington was bled and whether this procedure killed him. There is no doubt however that he was subject to very painful medical procedures that were common at the time. However, the medical procedures that were endured by George Washington were carefully recorded:

14th December 1799, 7.30am – 12–14 oz (355ml–414ml) blood removed by bloodletting,

14th December 1799, 8.30am – gargled with a tonic made of molasses, butter, and vinegar, which nearly choked him to death,

14th December 1799, 9.00am – applied blisters of cantharides (see next page) to his throat,

14th December 1799, 9.30am – 18oz (532ml) of blood removed by bloodletting,

14th December 1799, 11.00am – 18oz (532ml) of blood removed by bloodletting,

14th December 1799, noon – given an enema (see next page) and gargled with a mixture of sage tea and vinegar,

14th December 1799, pm – 32 oz (946ml) of blood removed by bloodletting,

14th December 1799, 8.00pm – blisters of cantharides applied to his feet, arms and legs, and a wheat poultice (see next page) applied to throat,

14th December 1799, 10.20pm – George Washington, the first President of the United States of America died. During his treatment 40% of his blood had been removed from his body.

It was probably the removal of a large volume of blood from his body that caused his death.

The medical assistant on his right of the photograph is holding in his hands a pump which would have been similar to the one inserted through Washington's anus and into the large intestine and a mixture of warm water and vinegar pumped into his body. This process is known as an enema.

A drawing of a woman undergoing bloodletting taken from a medical textbook of 1715. The skin and the blood vessel would have been opened using a fleam or bloodletting knife as shown above. There Is no real evidence that the practice of bloodletting had any advantageous effect on the patient. George Washington had 40% of his blood removed by bloodletting.

He also gargled with sugar, butter, and vinegar. Sage may also have been added. Gargles were and are still used to relieve sore throats. They are easily made, and the use of sage tea was common. Sage is known to have antibacterial and antiseptic properties.

George Washington was given a wheat poultice. Wheat would have been heated with hot water and then applied to the front of the neck and bandaged. The poultice draws out the dangerous infections and reduces the painful swellings of the sore throat.

Blister beetle – this beetle contains a substance known as cantharades which causes blistering when applied to the skin.

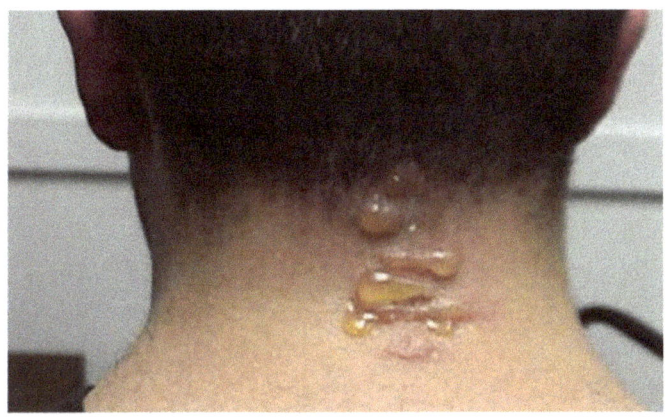

To remove even more fluid from George Washington he had crushed blister beetles applied to his throat and other parts of his body. The above picture shows blisters on the neck of a person who had a blister beetle squashed on his neck. In the treatment of George Washington, the blisters would have been cut open and the fluid removed which supposedly contained the dangerous chemicals that were infecting his body. There is no evidence that this procedure would have helped him to recover. There are over 7,500 species of blister beetle and they contain a substance known as cantharides which when applied to the skin causes severe blistering.

8

Arthur Coga – the Sheep Man

A blood transfusion is a lifesaving procedure where blood is transferred from one person to another. The transfusion is given for various medical conditions, e.g. during surgery, loss of blood in accidents, to improve the health of a person with a particular disease, cancer for example, or liver disease to name just a few. The procedure is very safe, but nevertheless it is essential that the person receiving the blood receives blood of the correct blood group. However, the first blood transfusions carried out over 300 years ago were very gruesome indeed.

1665 – The first blood transfusion was carried out by Richard Lower (1631–1691) in February 1665. In this gruesome experiment blood was transfused from one dog to another dog. In his first experiment he tried to transfer the blood from the vein of one dog to the vein of another dog, but it was not a successful transfer because a blood clot occurred in the silver tube connecting the two dogs together. Therefore, in his second experiment, which was more complicated, he connected the artery of one dog to the vein of the other dog. In this experiment he first removed most of the blood from the recipient dog, so much that the dog howled with pain and almost died. To replace the blood lost he used two donor dogs that were in turn connected to the recipient dog and replaced most of the lost blood. After removing the tubes, the recipient dog leapt off the table, ran to its owner and rolled about in the grass, with no after-effects whatsoever. The famous scientist Robert Boyle questioned the process. He wanted to know if when transfusing the blood in dogs the recipient of the blood changed into the breed of the donor dog!

1667 – In June 1667 Jean Baptiste Denis, a French doctor, was the first person to transfuse animal's blood into a human, a 15 year old boy who was suffering from a fever. Denis stated the boy was cured.

1667 – On November 23rd, 1667, Richard Lower transferred blood to a 22-year-old clergyman called Arthur Coga. About 12 oz (355ml) of blood was removed from a lamb and collected in a bowl. An incision was made in a vein in Coga's arm and about 7 oz (207ml) of blood removed. Then using a silver tube, 10oz (296ml) of the lamb's blood was transferred to Arthur. This was the first animal to human blood transfusion in England. Arthur was paid 20 shillings (about £180 in today's money) for taking part in the experiment. He took part in a second experiment and was paid the same amount of money. When asked if he would participate in another experiment he declined. After the experiments, Arthur started writing about his experiences using the name 'Coga the Sheep'. In his writings complained about the scientists who had changed him into a sheep, but the problem was that he could not find his fleece!! This claim stopped blood transfusions in England for over 150 years.

1818 – Doctor James Blundell carried out the first human to human blood transfusion.

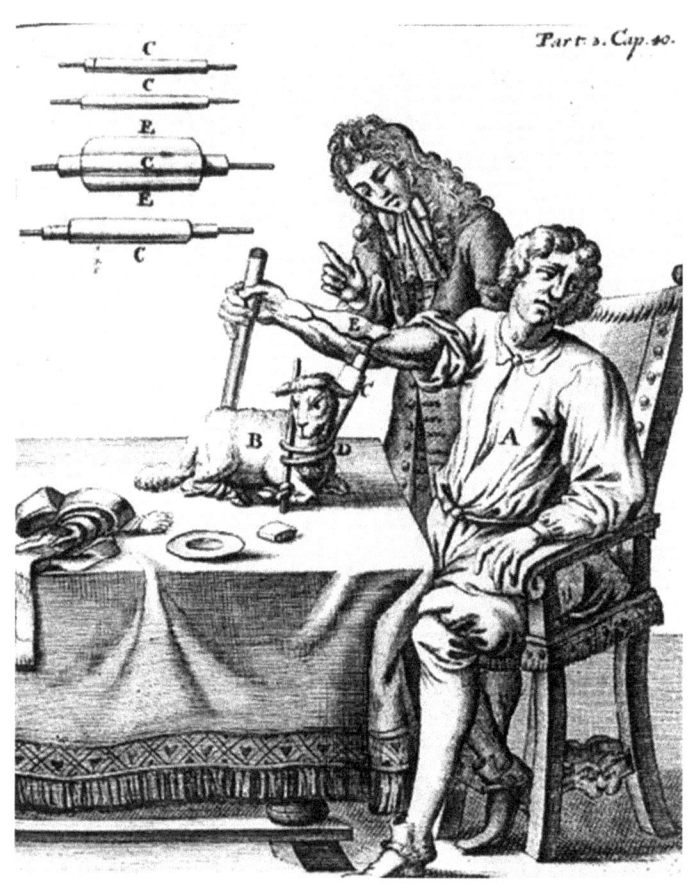

The drawing above shows the transfusion of a sheep's blood into a human. This transfusion was carried out by the English doctor Richard Lower in November 1667.

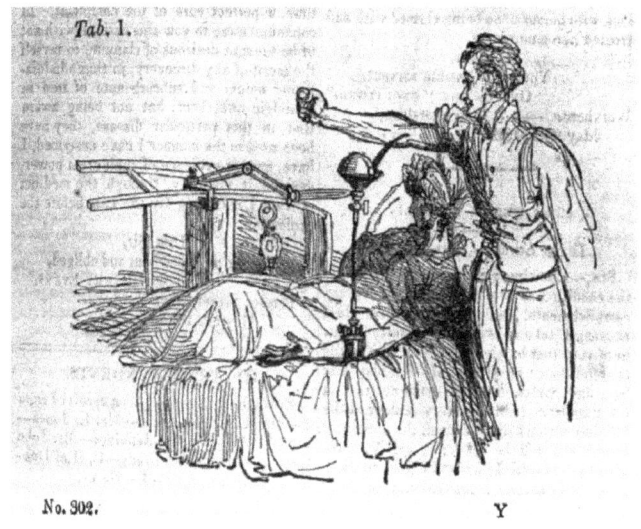

Tab. 1.

Dr James Blundell (1790–1878) was an English physician who was a pioneer in developing techniques for transfusing blood. In 1829 he successfully transfused 4 oz (120ml) blood from a husband to his wife when she had lost blood during the birth of their child. During his life Blundell performed ten human to human blood transfusions with only five being successful. He did not know that for blood transfusions to be successful the donor and the recipient had to have compatible blood groups which were discovered in 1900 by the Austrian scientist Karl Landsteiner. Blindell's successful blood transfusion were just a matter of luck!

The apparatus shown in the above drawing in which the husband is giving blood to his wife was developed by Blundell. So successful was his apparatus and medical practice that in his will he left £45 million pounds!

9

It's Out, it's Out!!

Isambard Kingdom Brunel (1806–1859) was one of the greatest ever engineers. In a poll carried out in 2002 he was voted as the second most famous Britain (in 1st place was Winston Churchill). Brunel was responsible for unique designs that revolutionised engineering. Among his greatest achievements were the design and building of the Clifton Suspension Bridge and SS Great Britain the first propellor driven ocean-going iron ship.

On the 3 April 1843, Brunel was entertaining his children with a magic trick. This involved hiding a gold half sovereign (a gold coin) under his tongue. Unfortunately, the trick went wrong, and he inhaled it, and it became stuck in his right bronchus (one of the tubes that leads into the lungs). The coin measured 2cm in diameter and weighed 3g. The inhaling of a coin is most unusual, swallowing is much more common. After Brunel inhaled the coin, he was to endure several weeks of pain and discomfort. Records exist of what happened to him.

3rd April 1843 – Brunel inhales coin. Back struck several times to remove coin, but with no success.

6th April 1843 – Brunel developed bad cough.

9th April 1843 – Brunel took aperient pills, but these were vomited back. Aperient pills reduce constipation.

11th April 1943 – Brunel again had a bad cough. Dr Brodie (Brunel's doctor) listened to Brunel's breathing, but there were no problems.

19th April 1843 – Brunel was bent over a chair with his head pointing downwards. He had a violent coughing fit and Brunel felt the coin move, but only further down into his lung!

20th April 1843 – Dr Brodie listened to his breathing again but there were no problems.

25th April 1843 – Brunel designed and built a hinged platform to which he was strapped and swung from side to side to dislodge coin but was not successful. His back was struck several times, but his coughing fit was so bad that the procedure was stopped.

27th April 1843 – Dr Brodie (and his colleagues Dr Key and Dr Hawkins) decide to cut open his throat and insert very long forceps designed by Brunel, into the right bronchus and remove the coin. There was no anaesthetic to reduce the pain. Brunel suffered a severe coughing fit, and the procedure was stopped.

2nd May 1843 – Long forceps again introduced into Brunel's throat and lung, but the procedure failed again.

3rd May 1843 – A decision was taken for Brunel to rest and recover his strength. A plug was placed into the cut in his throat to keep the cut open.

13th May 1843 – Brunel strapped to his platform again. His back was struck again, and the gold sovereign came out though his mouth.

20th May 1843 – Brunel fully recovered.

3rd June 1843 – Opening in Brunel's throat fully healed.

During April and May 1843, the talking point in Britain was the incident of Brunel and his inhaled gold coin. When it was known that the coin had left Brunel, people ran out into the streets and shouted, "It's out, it's out!"

Isambard Kingdom Brunel – one of the greatest engineers of all time.

Painting of the launch of SS Great Britain in Bristol in 1843, the first propellor driven ocean going iron ship – designed by Brunel.

The Clifton Suspension Bridge was originally designed by Brunel. The bridge was opened in 1864, five years after Brunel's death.

The top drawing shows one of the ways that was used to remove the gold coin – his back was slapped to try and dislodge the coin. The lower drawing shows the hinged platformed designed by Brunel which was successfully used to dislodge the coin which can be seen just below his head.

10

Maggots in the Body

You may find this horrible and revolting, but sometimes live maggots are placed in wounds to help them to heal. Maggots are part of the life history of a fly. War creates horrendous injuries to the body, they nearly always become badly infected and the first use of maggots was during war. In the early nineteenth century Napoleon's surgeon, Dominique-Jean Larrey, used maggots to clean wounds. Maggots were also use in the American Civil War 1861–1865. The first experiments and scientific studies in the use of maggots to clean wounds were investigated by an American surgeon called William Baer in 1929. He selected 21 patients who had open infected wounds and inserted maggots into the wounds every four days for up to six weeks. All his patients recovered and the wounds healed successfully. So successful was this therapy that in the 1930s over 300 hospitals in America were using maggots to cleanse wounds.

In the 1940s the antibiotic penicillin was discovered and the use of maggot therapy was dramatically reduced. Large open wounds are very distressing for patients and become infected with bacteria. Nowadays these bacteria are becoming more and more resistant to antibiotics and therefore there has been a return to maggot therapy. About 9,000 bags of maggots are used by the National Health Service every year.

Of course, any old maggots cannot be used. Some maggots feed on dead flesh, some on living flesh and some feed on both dead and living flesh. Research has shown that the most efficient maggots are those from the greenbottle fly, an insect that is common in Britain and in other parts of the world. The greenbottle flies are bred in very large sterile cages and the maggots produced that eat dead flesh, are packaged into small parcels each containing 20–100 maggots. The maggots, only a few millimetres in length, are inserted into the wound and the wound is then covered with a sterile dressing and allows oxygen into the wound. After three to four days the dressing is removed, and the maggots removed which have increased to about 12mm in length. The maggots only remove dead flesh and do not bury into the body. Using normal antibiotic treatment to clean wounds may take three months – the maggots can do the same job in days.

One of the problems with maggot therapy is that the patient can feel the maggots crawling about in the wound and this is sometimes very distressing for the patient. However, maggot therapy although horrible can be very effective, cheap, and surprisingly successful.

Greenbottle fly. It is the maggots from this fly which are used in maggot therapy.

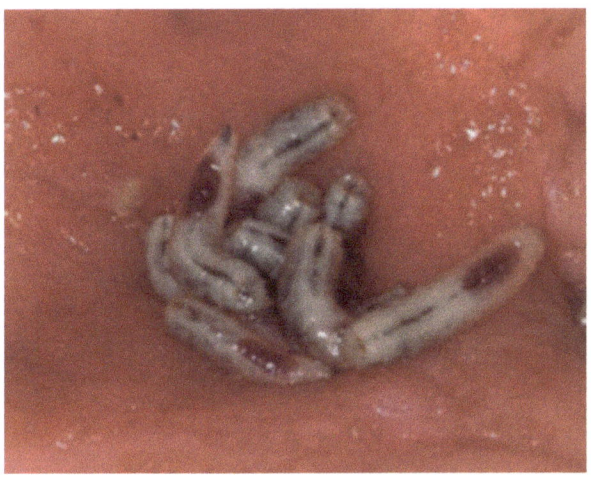

Maggots – a stage in the life history of flies. The maggots of different species of flies feed on dead flesh, living flesh or both.

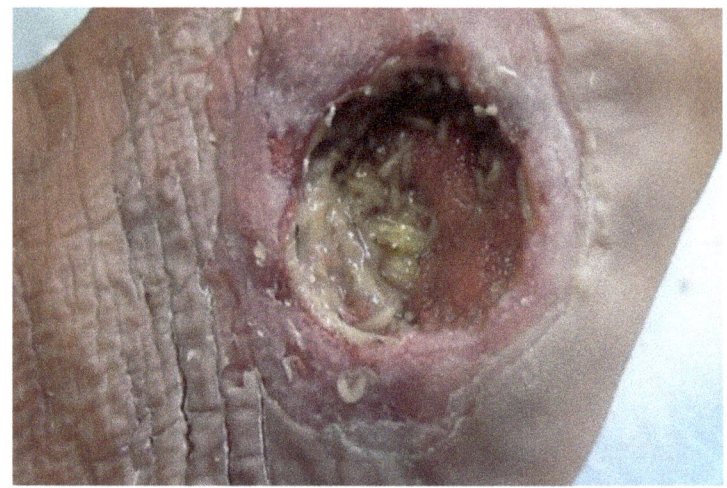

About 10% of people who suffer from diabetes will suffer from diabetic foot ulcers at some time during their life. An ulcer is an open wound and if foot ulcers are left untreated they can cause very serious problems for the diabetic. In the most serious cases, e.g. the toes or foot, amputation has to take place. Maggot therapy can be used to decrease the chance of amputation. In this photograph maggots are being used to remove the infected part of a foot in a diabetic foot ulcer. In England 9,000 amputations take place every year due to diabetic foot and toe ulcers.

11

Black Vomit and the Vomit-
Drinking Doctor

Suffering from a high temperature, bleeding from the mouth, the eyes and the nose, bleeding in the intestines, vomit black in colour because it contains blood – then probably you have yellow fever. Yellow fever is one of the most feared diseases and is still common in certain parts of the world, e.g., South America and Africa. It is caused by the Yellow Fever virus which is spread by an infected female mosquito when it bites humans. The disease has had devastating effects on populations. In 1793, 20,000 people died in the city of Philadelphia in America and 20,000 left the city to avoid the disease. The City Authorities encouraged people to keep apart, avoid shaking hands and cover their faces with handkerchiefs (does this remind you of a recent disease!) When Spanish troops occupied Cuba 16,000 died between 1895 to 1898. It is thought that even now 200,000 people suffer from Yellow Fever with as many as 60,000 deaths.

One of the first pioneers in trying to find out the cause of Yellow Fever was a third-year medical student called Stubbins Ffirth who was studying at the University of

Pennsylvania. He was absolutely convinced that the disease was caused by the high temperatures of summer which caused stress. He also stated that the disease was not contagious (passed from one person to another) and between 1802 and 1803 he carried out a series of horrible experiments on himself.

Experiment 1 – He made cuts in his arms and smeared his arms with the black vomit from a patient suffering from the disease.

Experiment 2 – He poured black vomit into his eyes.

Experiment 3 – He fried the black vomit and breathed in the fumes.

Experiment 4 – He drank the black vomit undiluted.

Experiment 5 – He smeared his body with the blood, saliva, and urine from an infected yellow fever patient.

He did not suffer the yellow fever disease from any of his experiments

However later investigations into his experiments concluded that Ffirth may have used the body fluids from a person who was recovering from the disease and therefore no longer infected with the disease. It was left to Carlos Finlay, a Cuban scientist to discover in 1881 that mosquitoes were responsible for the spread of this disease.

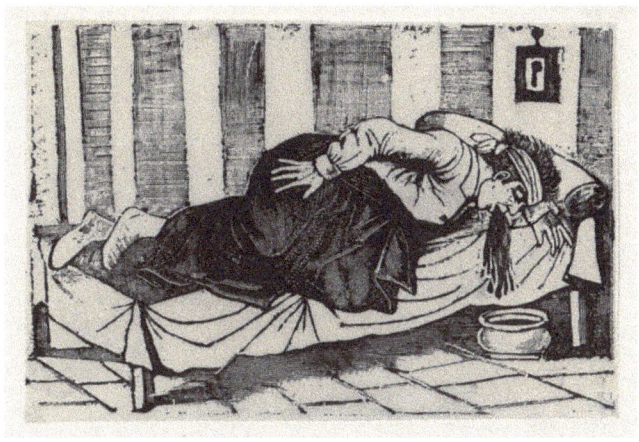

A drawing made in 1892 of the death of Aurelio Caballero from yellow fever. Note the black vomit which contains blood, one of the symptoms of the disease.

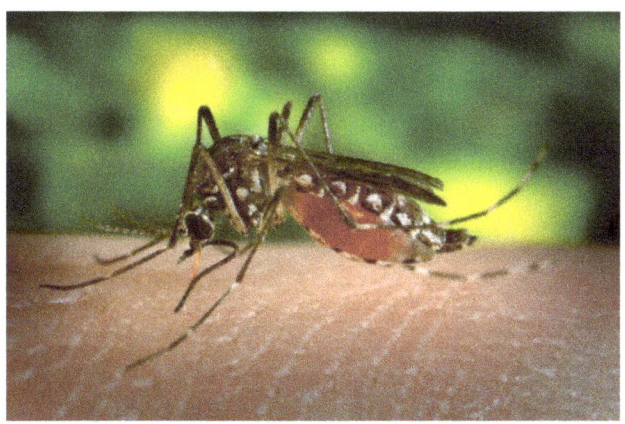

The female Yellow Fever mosquito which carries the virus and is responsible for Yellow Fever. If the female mosquito feeds on a person who has the disease and then feeds on another person, the virus is passed on.

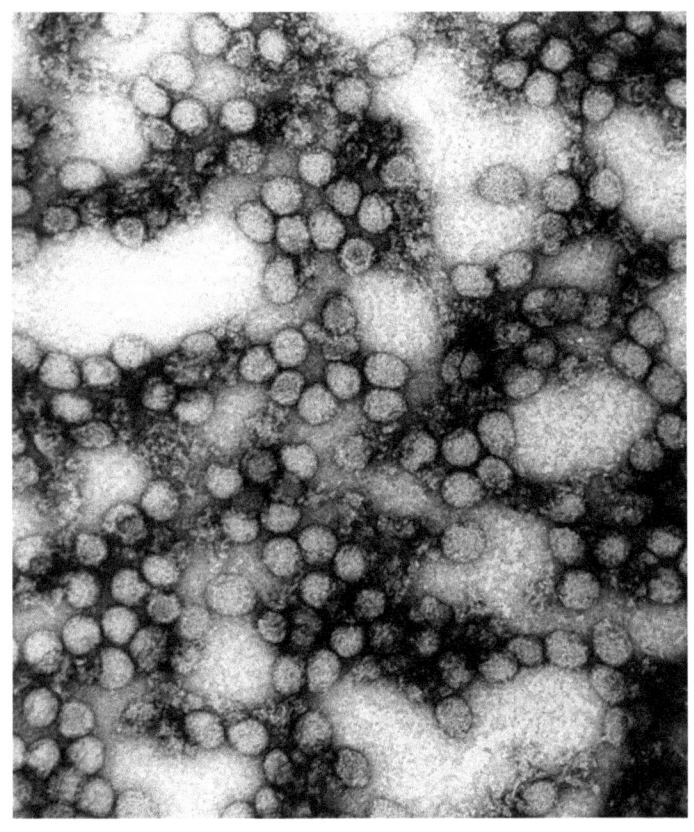

A photograph taken through an electron microscope showing dozens of the virus particles that cause Yellow Fever. It is difficult to Imagine that these particles that are only 50 millionths of a millimetre in diameter can cause such a terrifying disease.

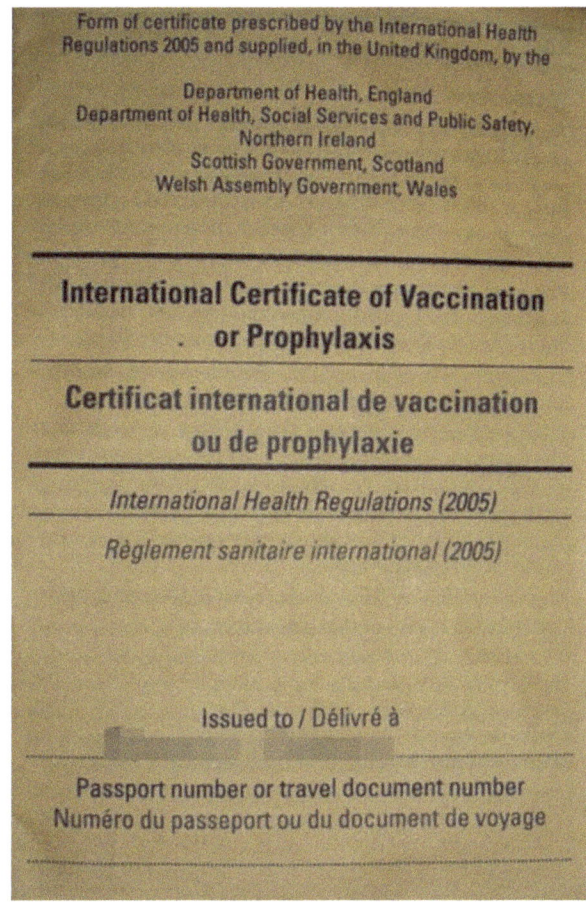

Form of certificate prescribed by the International Health Regulations 2005 and supplied, in the United Kingdom, by the

Department of Health, England
Department of Health, Social Services and Public Safety, Northern Ireland
Scottish Government, Scotland
Welsh Assembly Government, Wales

International Certificate of Vaccination . or Prophylaxis

Certificat international de vaccination ou de prophylaxie

International Health Regulations (2005)

Règlement sanitaire international (2005)

Issued to / Délivré à

Passport number or travel document number
Numéro du passeport ou du document de voyage

A vaccination is available to protect against the disease and a proof of vaccination certificate against Yellow Fever is required by many countries in Africa, Central and South America before you can enter the country. Vaccination is the most important means of preventing yellow fever and gives lifelong protection. At present there is no anti-viral drug available for this disease.

12

Plagues of Mice

On the 13th of May 1787 a fleet of eleven ships left Portsmouth harbour bound for Australia. The fleet consisted of two Royal Naval ships, three store ships and six convict ships. Some of the convicts had committed terrible crimes, others very minor (e.g. stealing a loaf of bread), but they were to be the very first settlers and formed the foundation of the Australian nation as we know it today. The living conditions for the convicts were atrocious – they lived in cramped, foul-smelling squalor and they were not even allowed on deck. They suffered from fleas, bed bugs, lice and cockroaches, and mice and rats were everywhere. After travelling 24, 000 km (15,000 miles) on a journey taking 250 days they landed in Botany Bay, New South Wales, Australia. Mice escaped from the ship. In the following years the mice would cause very serious problems for many areas of Australia.

For reasons that are not fully understood, populations of mice suddenly explode, and the number of mice reaches frightfully high numbers and are called plagues. Mice can breed when they are only six weeks old and have a litter every 21 days. A pair of mice can give rise to 500 offspring in a season! The first mouse plague occurred in South Australia in 1872.

One of the largest was in 1917 in the states of Queensland and Victoria. The mice ate their way through wheat crops – they ate boots, tablecloths curtains, clothes, books, they bit babies in their cots and destroyed telephone cables. They invaded a zoo and frightened lions and elephants; cats even lived up trees. People tried their best to kill the mice and 100 million mice were killed weighing 1,500 tonnes. In one night, 200,000 mice were caught and killed. In 1993 a plague of mice caused A$96 million of wheat growing in South Australia. In 2021 mice stripped a supermarket of all its food. Hundreds of mice were killed but the smell of mice urine and dead mice (from various poisons used to kill them) made the supermarket a complete disaster area. A mice plague in New South Wales has lasted for almost two years from 2019 to 2021. This has had devastating effects on communities in the area. Farmers have seen their crops eaten and their farming machinery disabled as the mice eat their way through electrical wiring. Such devastation and the mice invading properties has created many mental health problems. The noises made by the mice and the smell of rotting flesh has made sleep impossible in many cases. The rotting animals has been ideal for flies to lay their eggs and swarms of flies abound. The control of mice plagues is very difficult. Poisons are available and are used, but they can also kill wildlife that naturally feed on the mice, e.g., nankeen kestrels, black shouldered kites, and barn owls.

500,000 mice caught and killed in Lascelles, Victoria, Australia, in May 1917.

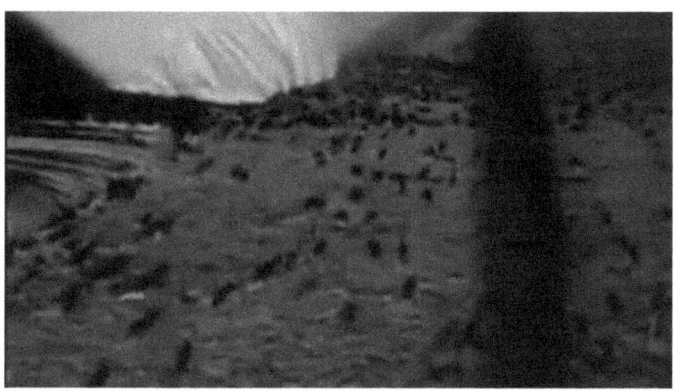

The bed of a farmer in Australia that has been invaded by mice during a mice plague.

The four photographs above show the large number of mice killed on one farm, in Australia in 2021. The Australian government made an additional A$ 100 million (£80,000,000) to combat a mouse plague in New South Wales, Australia in 2021.

It is hard to believe that the cute little animal shown above can bring devastation to many parts of Australia. Also shown above is the destruction of a farmer's store of hay by mice.

13

Toxocariasis

The National Health Service and Councils often produce information sheets to help people understand health issues. The Mid and East Antrim Borough Council in Northern Ireland produces an information sheet (see page 71) about a parasitic worm called Toxocara that can cause problems, sometimes very serious in humans.

In most cases toxocariasis is not serious and many people especially adults may not notice any symptoms of the disease at all. Severe cases are rare in the United Kingdom, but they do occur in young children. The worm lives in dogs and as many as 200,000 worm eggs per day can be found in dog faeces (poo). If these eggs are transferred to the human mouth and swallowed the eggs turn into larvae which migrate to the liver, heart, lungs, brain, and eyes through the blood circulatory system. It is when the larvae reach the eyes that serious problems occur. The parasite was first discovered in 1956 and is present in over 100 countries including the United Kingdom, France, India, Japan, USA, Brazil, and Australia. In the USA there are 100,000 cases of toxocariasis a year with 10,000 cases involving the eyes, with 700 million children becoming blind. In the United Kingdom 24% of all dogs and

33% of all cats have these parasitic worms. Our immune system prevents serious symptoms from developing in most people.

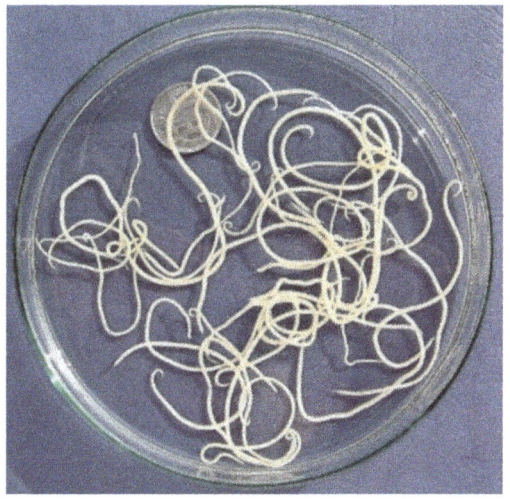

Dogs and cats may have parasitic worms. The worms lay eggs inside the dogs and cats, and these are found in the animal's faeces (poo). If the eggs are swallowed by humans a disease called toxocariasis can occur. The Toxocara worms shown above were found in dog faeces. The male worms are about 4cm–6cm in length and the females slightly longer at 6cm–10cm in length. The coin in the photograph is to give some idea of the length of the worms for comparison. One of the reasons for the parasite not being common in this country is that puppies are nearly always given medication by the vet to kill these parasitic worms, a process known as de-worming.

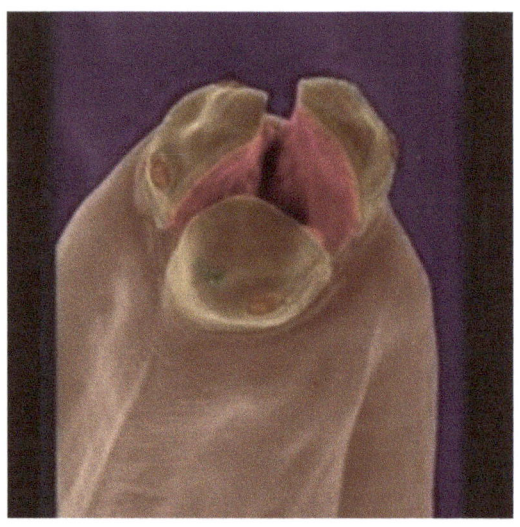

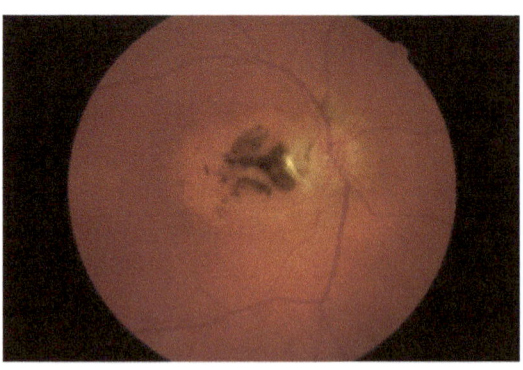

The top photograph, on page 70, shows the head of head of the Toxocara worm. At the bottom is a photograph of the retina at the back of the eye. The dark area in the centre of the retina shows the larvae of the worms which could eventually cause blindness in the infected person.

Toxocariasis Information Sheet

Dog faeces can pose a significant health risk to humans, particularly young children as their immune systems are not fully developed. Children are also more likely to come into contact with soil or sand that contains dog faeces whilst playing in parks, gardens and playgrounds. All faeces contain bacteria that can cause stomach upsets, but the greatest risk is from toxocariasis.

Toxocariasis is particularly hazardous to small children as it can result in blindness. You might have heard of Toxocariasis, but do you know exactly what it is and how to prevent it? Even if you don't have a pet, make sure you're aware of the risks.

What is Toxocariasis?

Toxocara is the name given to a species of roundworm commonly found in dogs and cats. Virtually all puppies are born with Toxocara. Puppies and kittens can also be infected with Toxocara through their mother's milk or from environmental contamination. The type of toxocara found in dogs can endanger human health. The type of toxocara found in cats has only rarely been associated with cases of Toxocariasis (Toxocara infection).

How is Toxocariasis spread?

Microscopic toxocara eggs are present in the faeces of infected animals. These eggs have thick, sticky shells which means that they can remain infective in the soil for two to four years after the faeces have disappeared. The sticky shell helps eggs to adhere to fingers or clothing.

How do humans become infected with Toxocariasis?

By accidentally swallowing the infective Toxocara eggs. This is why crawling babies and toddlers are most at risk; they tend to put dirty fingers and toys into their mouths. Medical records show that approximately 100 new cases of Toxocariasis are diagnosed each year.

What happens once humans get infected?

Once swallowed, Toxocara eggs release larvae into the intestine. These larvae travel through the body until they die, which may take several years. The symptoms of this disease can be unpleasant and difficult to treat. They can include stomach upset and pain, headache, sore throat, wheezing and listlessness. In some cases, larvae reach the eyes where they can cause sight problems and in some cases blindness.

A 10 step guide to preventing Toxocara

1. NEVER take a dog or cat to a play or sports area.
2. Always clear up pet faeces immediately using a nappy sack, a carrier bag or a poop scoop bag, and deposit it in the nearest dog bin or, if at home, in a safe and secure bin.
3. Make sure that children always wash their hands, especially after playing outside and before eating.
4. Discourage your child from sucking its fingers. This is one of the most common ways for children to get infected.
5. Don't allow children to play on the ground and eat using their fingers at the same time.
6. Be sure that floors where children play at home are cleaned with antibacterial cleaners. This practice kills Toxocara eggs that may have been transported into your home on your shoes or on your pet's paws.
7. Don't put small children on the entrance floors of public buildings.
8. Don't leave young children unsupervised with a pet.
9. Fence around play areas to keep your children safe and also to keep dogs away from play equipment.
10. Cover sandpits to keep cats and dogs out.

www.midandeastantrim.gov.uk

Mid & East
Antrim
Borough Council

14

When Biologists Get it Wrong!

102 Cane Toads were introduced by biologists into Australia in 1935 to control an insect pest of sugar cane. This method of using one species to control another species is called biological control. The use of the cane toad had been very successful in controlling the sugar cane beetle in Puerto Rico (an island in the Caribbean) in the 1920s. It therefore seemed a good idea to introduce them into Australia to control the sugar cane pest called the Grey Back Cane Beetle that was having a disastrous effect on the crop in this country. However, this has not been successful in Australia and the cane toad has been responsible for the severe reduction in the populations of many native species of animals. There are thought to be at least 200 million cane toads in the country, but some estimate the figure to be much higher at 1.5 billion.

Cane toads are voracious feeders and eat insects, frogs, lizards, mammals, birds and even their own tadpoles. Any living organism that can fit into its mouth can be eaten. They are also very large about 22.5cm (9 inches) and weigh 2kg (4 lbs) and they live for 10-15 years. The female cane toad can lay 8,000 to 20,000 eggs in a long 'string' 20m in length.

When the eggs are fertilised by the male toad very large number of tadpoles are produced.

When threatened glands on the back of the cane toads release a milky coloured very toxic fluid from the skin called bufotoxin. There is sufficient bufotoxin in the animal to kill a 3m long crocodile! In the Northern Territory of Australia there has been a 75% reduction in the population of freshwater crocodiles because young crocodiles feed on the toads. It is thought that 75 species of animal are at risk from the toad including several species that are endangered (this means a serious risk of extinction) including the Northern Quoll which feeds on the cane toad, although environmental factors such as fire has also reduced the population of this rather cute animal. There are other animals that feed on cane toads, but they have unusual feeding strategies to overcome being poisoned. The rakali (a species of water rat) flips the toad onto its back and then cuts through the belly and eats the non-toxic liver and heart of the toad. Two birds of prey, the Black Kite, and the Whistling Kite, only eat the non-toxic tongues of cane toads that have been killed on roads.

The cane toad has not been particularly successful in controlling the grey back cane beetle. The beetle feeds at the top of the sugar cane and the toad cannot reach them or jump high enough to feed on them! It was much easier for the toad to catch other prey.

The grey back cane beetle is a major pest of sugar cane.

The cane toad was introduced into Australia to control the grey back cane beetle. Unfortunately, it was not particularly successful.

Cane toads are large, weighing 2kg and reaching 23cm in length.

The spawn "string" of the cane toad. It can reach 20m in length and contain up to 20,000 eggs.

Cane toads are voracious feeders even eating their own species – cannibalism! They will also feed on mice and on birds.

This rather cute Northern Quoll is a marsupial (a pouched mammal, related to the kangaroo) and feeds on cane toads. Unfortunately, this has led to a reduction in the population. One of the reasons is the poison produced by the toad.

The Rakali or water rat is about the size of a rabbit and feeds on fish and the poisonous cane toad. It has an unusual strategy for eating the cane toad. The rakali flips the toad onto its back and then cuts into its belly and feeds on the heart and liver of the cane toad, thus avoiding eating the poisonous skin of the toad.

15

The Bombardier or Farting Beetle

Some animals will go to extraordinary lengths to defend themselves against predators. One group of animals who defend themselves in a most unusual way is the bombardier or farting beetles which are only about 2.5cm in length. There are over 350,000 different species of beetle (some experts believe the number is nearer to 1,500,000) with about 500 different species of bombardier beetles, of which only two are found in the United Kingdom and they are rare with one described as the rarest insect in the United Kingdom.

They are called farting beetles because a very hot, horrible smelling liquid/gas mixture is expelled from the end of the beetle's abdomen when they are threatened. There has been much research carried out to find how this 'farting' takes place. Inside the abdomen of the beetle are two small sacs (see the drawing on the opposite page), one containing the chemical hydrogen peroxide, the other a chemical called hydroquinone. When the beetle is threatened these two chemicals are mixed and another chemical called a catalyst is added. An explosive chemical reaction takes place inside the explosion chamber and a mixture of gases and liquid leaves the abdomen at a temperature of 100°C (212°F) and a speed of 35 km/h (22mph). This strategy is very effective in protecting the bombardier beetle against predators such as ants, birds, spiders, frogs, and toads.

What is even more remarkable is that the beetle can use this strategy after it has been swallowed! The Japanese Tree Frog is a predator of many beetles including the bombardier beetle. The frog uses its sticky tongue to capture and then swallow the beetle. Once inside the stomach the beetle then releases the explosive high temperature chemical mixture which so irritates the stomach of the frog that it vomits the bombardier beetle out of its mouth. The beetle is not harmed, and it can be in the stomach for several minutes and up to one hour before it uses this bizarre strategy to escape from the frog.

Charles Darwin was an enthusiastic collector of beetles. On 17th October 1846 he wrote a letter to another naturalist called Leonard Jenyns about an incident that had occurred when he was collecting beetles on the banks of the River Cam near Cambridge. He was looking under some leaves when he saw an interesting beetle which he collected with his hand. He then saw another beetle which he collected with his other hand. He then observed another beetle which he wanted to collect. Not knowing what to do, he writes "I gently squeezed one of the beetles by the teeth when to my unimaginable disgust and pain the little inconsiderate beast squirted acid down my throat." Darwin spat out the bombardier or farting beetle onto the ground and it ran away.

The bombardier or farting beetle. There are about 500 species of this beetle, two of which live in the United Kingdom, but are rare.

The bombardier beetle feels it is threatened because a pair of tweezers (simulating a predator) is holding one of its legs. The beetle can direct its boiling gas/liquid mixture towards the tweezers and so defend itself.

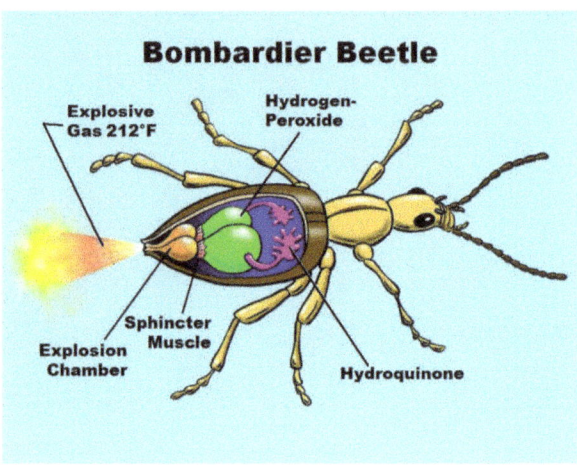

A diagram showing the position of the two sacs that hold hydrogen peroxide and hydroquinone. The sphincter muscle controls the amount of the two chemicals reaching the explosion chamber. After they are mixed with the catalyst the resulting gas/liquid leaves the abdomen at a temperature of 100°C (212°F).

The boiling gas/liquid leaving the bombardier beetle.

The hot chemicals at 100°C are expelled from the end of the abdomen of the farting beetle. This method is an effective method of deterring predators.

16

The Kissing Bug

Fast asleep in Central and South America and without a protective net you could be bitten by the kissing bug. This bug bites you on the lips, hence the name kissing bug. This 2.5cm (1 inch) bug carries a microscopic parasite that causes Chagas disease that infects six million people and causes 10,000 deaths a year. The disease is common in Central and South America. The Brazilian parasitologist Carlos Chagas (1879–1934) discovered the disease in northern Brazil in 1909. Although there are about 150 species of the kissing bug or assassin bug as it is sometimes called, there are only five species that are thought to be important in spreading Chagas disease in humans.

During the day the kissing bug hides in cracks in houses, under sinks and in bedrooms. The bug is nocturnal and is attracted by high concentration of carbon dioxide which is found close to the face of the intended victim. The bug has a very sharp pointed mouthpart called a proboscis which is used to pierce the skin of the lip or eyelid of the sleeping victim. Saliva is now injected which numbs the area around the bite, prevents clotting of the blood and increases blood flow to the area. The kissing bug now feeds on blood for about 30 minutes. During feeding the bug deposits its poo onto the skin. If the poo contains the parasite, then there is the possibility of the person developing Chagas disease. All that is required is for the person to scratch the area around the bite and for the poo to enter the body through the scratch. It is the poo that contains the parasite and not the saliva. Strangely the disease takes many years to develop, and most people do not know they have the disease. However, about 45% of the people who suffer from Chagas disease develop heart disease and heart failure 10–30 years after being first bitten. Animals can also suffer from the disease including cats, dogs, rats and in particular armadillos. It costs about $1,000 a year for medical care for each person with Chagas disease, $11,600 for lifetime care. Globally $627 million is spent on medical care for people suffering from the disease.

The famous biologist Charles Darwin (1809–1882) is supposed to have died from Chagas disease. In his journal dated 26th March 1835 (when he was in South America) he wrote "It is most disgusting to feel soft wingless insects about an inch long [2.5cm] crawling over one's body; before sucking they are quite thin, but afterwards round and bloated with blood and in this state, they are easily squashed." Although Darwin was diagnosed with heart disease it cannot be proven he had Chagas disease because it had not been identified at the time of his death, but he had many of the symptoms of the disease.

The kissing or assassin bug, the insect which carries the microscopic parasite causing Chagas disease that causes 10,000 deaths a year. Notice the long sharp proboscis underneath the head that is pushed through the skin of the victim so the insect can feed on blood.

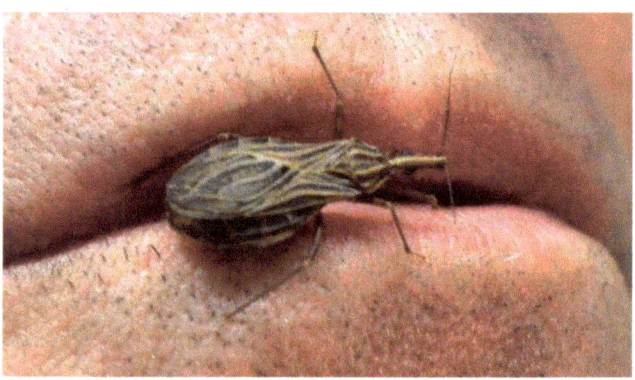

The kissing bug on the lips of a person. The bug has been attracted to the face by the high concentration of carbon dioxide from the person's breath.

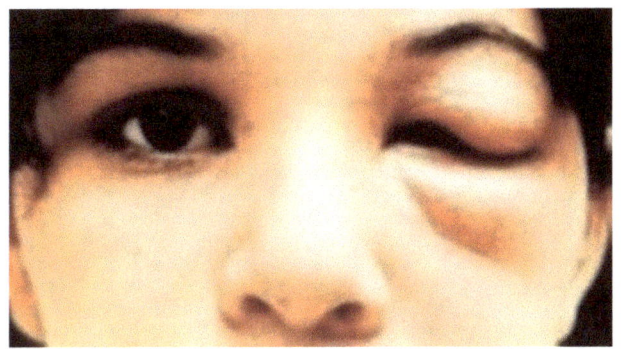

This person has been bitten on the eyelid by the kissing bug.

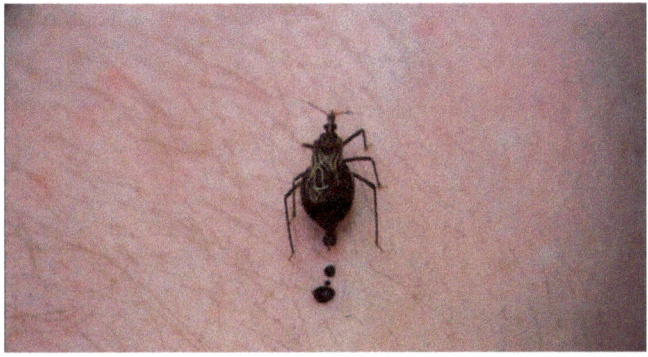

The dark red droplets behind the kissing bug is the poo (faeces) of the insect. If the insect is infected with the parasite, it will be found in the faeces and if the skin is scratched and the faeces enters the scratch or if the faeces enters the mouth then the parasite enters into the victim and will suffer from Chagas disease.

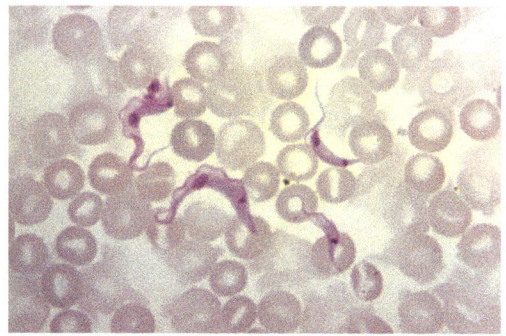

A photograph taken through the microscope of the blood of a person suffering from Chagas disease. The round structures are the red blood cells of the person. The long C-shaped structures (which have been dyed to make them easier to observe) are the parasites that cause the disease. It is difficult to believe these parasites that are about 0.04mm in length can cause disease.

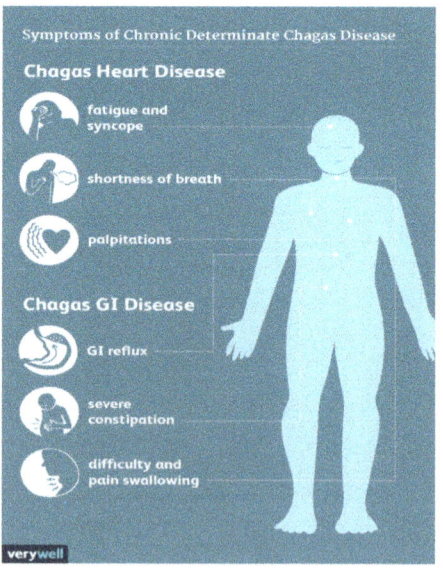

In America much information is made available to the public about the dangers of the disease. Syncope or fainting is caused by sudden drop in blood pressure. GI reflux or heartburn caused by the acidic food in the stomach coming back up the oesophagus or food pipe.

17

The Hairy or Horror Frog

There are over 8,000 different species of amphibians (this includes frogs, toads, newts, and salamanders) They range in size from the giant salamander that is over 150 cm in length and weighs in at over 50kg to the smallest frog and vertebrate (animals that have backbones) in the world. This unbelievable small frog is only 7.7 mm in length! It was discovered in Papua New Guinea in 2009 and weighs only 10mg. The frog is very unusual, not because of its small size, but because it does not have tadpoles in its life history. When the fertilised eggs of this frog hatch even smaller frogs (known as hoppers) emerge!

There are some very unusual amphibians and one of them is shown on the next page. In the photograph the frog known as the hairy or horror frog appears to have hairs on its back and legs. If this was true, the animal would be a mammal because only mammals have hairs; this animal is an amphibian. The "hairs" are not real hairs, but thin extensions of the skin that contain blood vessels which increases the area for absorbing oxygen. Only the male frogs have these "hairs." During the breeding season the male frog spends long periods of time under water protecting the fertilised eggs until they turn into tadpoles. The frog has quite small lungs and absorbs oxygen through its skin.

This hairy frog has one of the most bizarre methods of defending itself when threatened. When the frog senses danger it deliberately detaches bones in its hind legs and then forces the bones through the skin so that they become very sharp claws. Once the danger is over the claws are retracted back into the foot and new skin grows to cover the damaged skin. It is thought that the claws could also be used for helping them to climb over rocks. The hairy frog is the only vertebrate (animals that have a backbone) which has claws made of bone. This 10cm frog lives in Central Africa, is nocturnal, lives in forests and on agricultural land, and feeds on insects, spiders, slugs, and snails. Like all frogs once the food is in its mouth it pushes the food down the throat by using the back of its eyeball. It does not drink but absorbs water through the skin.

The Giant salamander, the world's largest amphibian, found in China. It can grow to over 150 cm in length.

The world's smallest amphibian, only 7.7mm in length. It is found in Papua New Guinea.

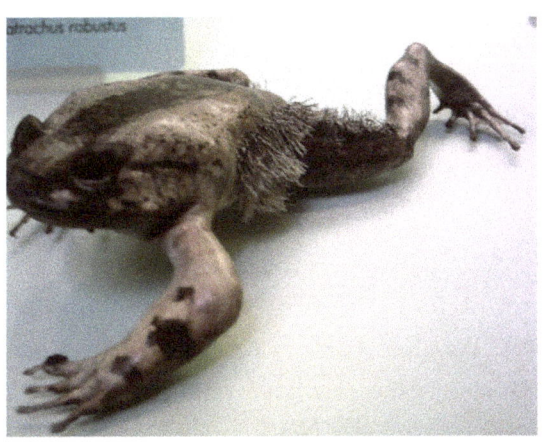

The Hairy frog with "hairs" on its back and thighs. The hairs are not real, but skin extensions containing blood vessels which increase the surface area for the absorption of oxygen.

The white claws are produced when the hairy frog is threatened. Bones in the toes are deliberately broken by the frog and are then forced through the skin to form claws. Once the threat is over the claws are retracted into the toes and become attached to other bones in the foot. The broken skin where then claws protruded is repaired by the frog.

18

Hunger and Balloon Swallowing

Most of us have felt hungry at one time or another. We have a kind of clock which is used in our daily lives which tells us when to sleep and when to eat. When it is dinnertime, we feel hungry. We can feel hungry when we smell chips cooking or when bread is baking. Even the colour of food can make us want to eat – a yellow banana seems more edible than a painted red one. Some people like certain foods so much they eat them when not hungry. However very few of us have really felt severe hunger when a dull ache begins in the stomach area of the body and becomes more and more painful often accompanied with severe headaches and fainting. The pain sensations are intermittent and appear and disappear many times an hour. Feeling hungry tells us that our bodies require food. Biologists have often asked the question "How do we know we are hungry?"

In 1911 a series of very famous experiments were carried out by two American doctors called Arthur Washburn and William Cannon to investigate "How do we know we are hungry." Washburn trained himself to swallow an 8cm diameter balloon with a tube attached – this is a very difficult and horrible procedure because swallowing a balloon causes the person to choke. It took every day for several weeks for Washburn to become accustomed to the balloon swallowing and not to feel he was about to choke. When he had become accustomed to the balloon being in the stomach for two or three hours it was filled with air. When the stomach muscles contract they would press against the balloon and the changes in pressure inside the balloon could be recorded. The simple diagram on the next page shows how the recordings were made. On the days the experiments were undertaken Washburn would have no breakfast or lunch and would arrive in the laboratory at about 2pm.

These experiments resulted in Washburn and Cannon producing their 'stomach contraction theory'. They found out that when we are hungry our stomachs contract. When the balloon in the stomach was inflated Washburn did not feel hungry, but when the balloon was deflated, and his stomach was empty the stomach muscles contracted, and he felt hungry.

It is over one hundred years since Washburn and Cannon carried out their rather horrible experiments and it is now known that feeling hungry is far more complicated than their theory. Although the stomach muscles do contract when hungry, it also involves chemicals known as hormones and strangely people who have had their stomachs removed because of disease still can still feel hungry!

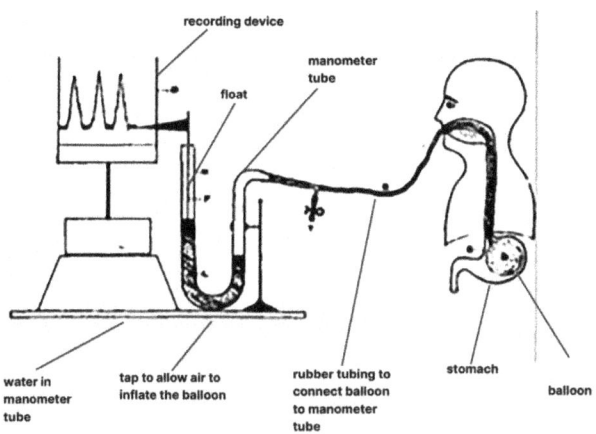

The diagram above is taken from a book written in 1916 by Anton Carlson and called 'The Control of Hunger in health and Disease'. The diagram shows how the balloon in Arthur Washburn's stomach was used to investigate the muscle contractions of the stomach and how they were recorded. (Labels have been added to make the diagram clearer). When the stomach muscles contract air is forced along the rubber tubing which causes the float to rise, and this is recorded on the recording device.

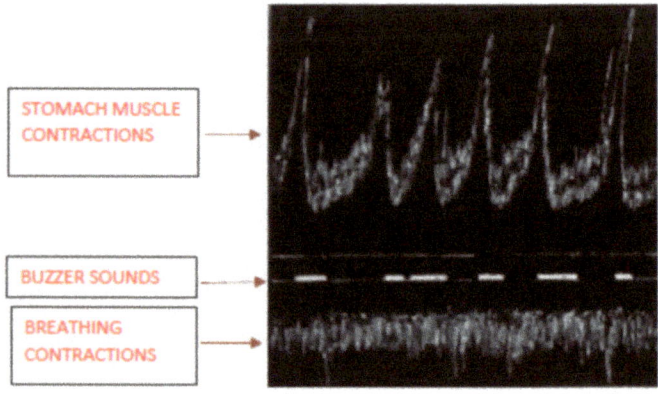

STOMACH MUSCLE CONTRACTIONS

BUZZER SOUNDS

BREATHING CONTRACTIONS

The recordings were taken over a period of ten minutes. During this time the stomach muscles contracted six times. The diagram above shows some of the recordings taken when Washburn was used in the investigation. The diagram also shows the breathing contractions of the muscles involved in breathing.

19

Jigger Fleas and Itchy Feet

Of the 2,500 species of flea, the jigger flea (sometimes called the sand flea) is the smallest at only 1mm in length. The flea lives in the sands on the coasts of Central America, the Caribbean and Africa. When a person with bare feet walks through the sand the jigger flea attaches itself to their skin. The flea now burrows into the skin and feeds on the blood of its host. As it feeds, the flea can become over 1cm in size. It is only the female flea that feeds in this way because it requires blood for the development of its eggs. After it burrows into the skin, the only part of the jigger flea that can be seen is the end of the abdomen which appears as a black dot. The flea breathes through the end of its abdomen. Although it is mainly the feet that become infected, other parts of the body may also be infected by laying in the sand.

When the feet become infected with the jigger flea, severe itching takes place with intense pain and great difficulty in walking. However, the flea contains many very nasty bacteria which cause horrible problems for the infected person. These infections include gangrene (the skin and underlying tissues die due to the lack of blood supply and become black and rot) and tetanus (affects the nervous system and causes muscle contractions of the jaw and neck and interferes with breathing). These infections become even worse when the jigger flea dies in the skin. Many people with jigger fleas have their toes amputated.

The disease originated in the West Indies but spread to other countries in the seventeenth and eighteenth centuries when trade ships and expeditions travelled the world, particularly from the West Indies to Africa, and sailors became infected after walking barefoot on sand. It is also known that when Christopher Colombus' ship the Santa Maria was shipwrecked on 24th December 1492 in the Caribbean the sailors became infected with jigger fleas. Research carried out in 2019 on the coast of East Africa showed that 48% of schoolchildren were infected with the jigger flea. In some areas of Brazil over half the population is infected. Application of coconut oil to the feet stops the flea from penetrating the skin. At the present time there is no effective treatment for the infection. Using sharp unsterilised objects to remove the fleas from the skin, as is traditionally used, makes the situation worse causing even more bacterial infection.

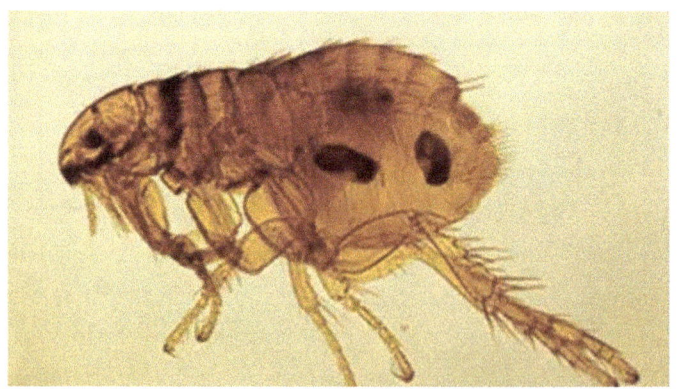

Although the jigger flea shown above is less than 1mm in length, it can inflict very serious damage to the feet as in shown in the photographs below. The black dots on the feet indicates the end of the abdomen of the dead jigger fleas.

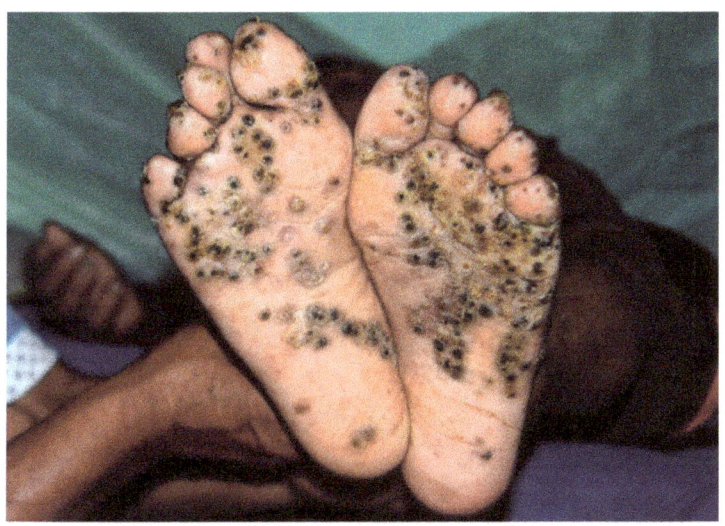

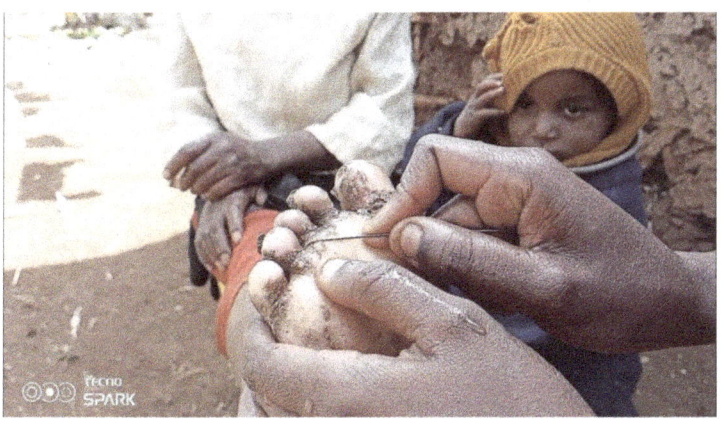

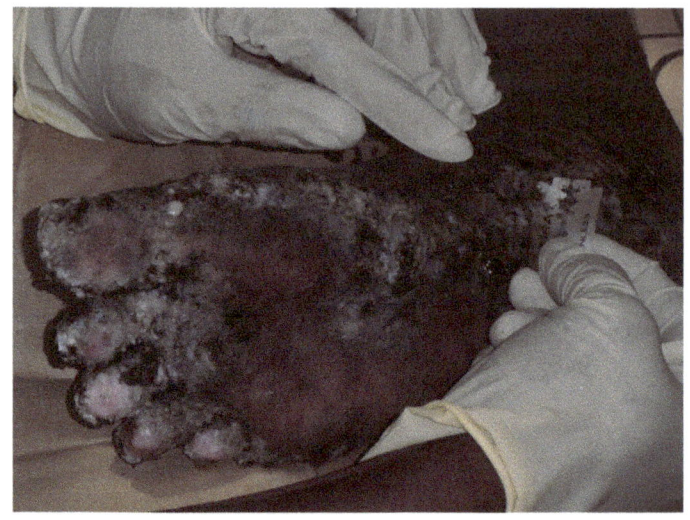

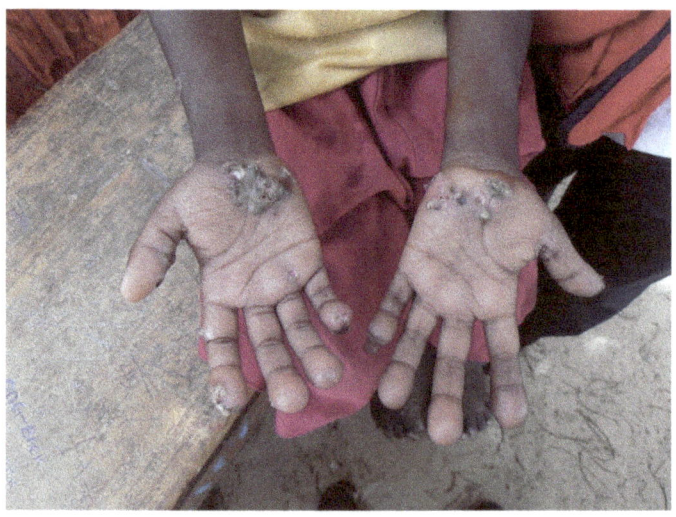

The feet and sometimes the hands can become infected with parasitic jigger flea. The only way to remove the fleas is with a sharp scalpel or even a razor blade. Over 2 million people in the world are infected with the flea.

20

Ophidiophobia

If you were to choose a group of animals that would create more fear and disgust than any other animals, it would have to be snakes. About a third of all adults have a great fear of snakes – a condition known as ophidiophobia. Snakes are reptiles and have a dry, scaly skin, a heart with three chambers (we have four) and one lung (we have two). There are about 3,900 species of snake and of these about 400 are venomous and fewer than this are deadly to humans, but over 100,000 people a year are killed by venomous snakes. All snakes are carnivores – there are no vegetarian snakes. If a snake is venomous, it kills its prey by biting and injecting toxins. If a snake is poisonous, it does not bite and inject toxins, but its toxins are to be found throughout its body. If the snake is eaten, then you will ingest these toxins which can cause much suffering and even death. The best examples of poisonous snakes are the garter snakes of which there are 35 species.

The Reticulated python of Asia is the longest snake in the world and can easily reach 6m (20 feet) in length with some almost reaching 8m. They often lay in ambush to capture their prey of rodents, deer, and pigs which they crush to death before eating. The female lays between 25–50 eggs and when the eggs hatch the baby pythons are about 60cm (2 feet) in length.

Discovered in 2008, the Barbados Thread snake (shown above) is the world's smallest at only 10cm in length. Looking more like an earthworm than a snake, the female lays only one egg which is about the size of a grain of rice. The snake eats ants and very small grubs.

The adder (shown above) is the only venomous snake to be found in the United Kingdom. Reaching 60 cm in length it is found throughout Europe. It feeds on frogs, small birds, mice, and rats. The female does not lay eggs but produces live young.

Although only about 2m (6 feet) in length, the black mamba shown on the previous page (so called because the inside of its mouth is black) is one of the most dangerous snakes in the world. Its venom is highly toxic, and the snake is lightning fast and can move along the ground at 20km/hour. It is very aggressive not allowing humans to approach within 40 metres! Its bite is nearly almost fatal unless antivenom is available. It is the longest venomous snake in Africa.

The bazaar 50–90cm Tentacled snake lives in muddy lakes and paddy fields of Thailand, Viet Nam and Cambodia. Its two tentacles are unique among snakes and are thought to detect movement in the water They feed only on fish.

Garter snakes are very common in America and are non-venomous and shy. However, they are poisonous. They feed on frogs, newts, slugs, snails, and worms.

The Belcher sea snake is arguably the world's most venomous snake. They hardly bite and when they do bite humans they only inject very small amounts of venom. At about one metre in length, they inhabit the Indian Ocean and the northern seas off Australia. A few mg of venom can kill 100 humans.

21

Venomous Snakes

Just because some snakes are highly venomous it does not mean they kill the most people. There are several reasons for this. Some of these snakes are to be found in very remote areas where there is little human habitation; some snakes are shy and retiring and move away when approached by humans. It also depends upon the amount of venom injected into the person and some venomous snakes bite multiple times.

Australia has some of the most venomous snakes in the world and in the last ten years only ten people have been killed! Australians have learned to stay away from snakes. The inland taipan is the most venomous of all snakes and lives in the most remote areas of Australia. Some highly venomous snakes are extremely aggressive and are quite prepared to attack humans, eg black mamba.

The saw scaled viper of Africa, Asia and Sri Lanka kills more people than any other snake because it lives in densely populated areas, and it often encounters humans. The venom of this viper is not the most lethal, but the snake is aggressive and well camouflaged. About 50,000 people are bitten by this snake every year with 5,000 deaths.

It is thought that about 5,400,000 people are bitten every year by snakes with 100,000 deaths and 300,000 people being severely disabled with many of these having amputations caused by the toxins in the snake bite. Snake bite toxins contain some very nasty chemicals with venomous snakes having different combinations of these chemicals. Some snake bites cause the victim's blood to clot; some cause blood to leak from the blood vessels of the victim; some affect muscles, particularly those involved in breathing; some affect the nervous system, or it may be a combination of any of these. The effects of the snake bite also depend upon the amount of venom injected and the age of the victim.

A major reason why deaths from snake bites has been reduced in some countries is the development of anti-snake venom. In the preparation of the anti-snake venom a snake is "milked" of its venom and a very small amount is injected into a horse. The horse is not affected but it makes chemicals called antibodies to protect itself against the snake venom. After about 8–10 weeks blood is removed from the horse. The blood is now filtered in a very complex process to remove the antibodies which can now be injected into the victim. A whole range of anti-snake venom for the different venomous snakes is available in hospitals if the victim can be taken to the hospital in time. To make a pint of cobra anti-venom takes 69,000 "milkings" of cobra snakes.

The Inland Taipan snake – at only 1.8m – 2.5m in length with fangs 3mm – 5mm long this eastern and central Australian snake is the world's most venomous snake. One bite contains enough toxin to kill 100 people.

The India Cobra is famous for its hood which is raised if it is threatened. Up to 1.5m in length, but occasionally longer, is found in India, Sri Lanka, and Bangladesh. Responsible for many bites, but if antivenom available then lives can be saved.

The Saw Scaled viper is found in Africa, Asia and is responsible for more deaths than any other snake. Lives in densely populated areas.

The Fer de Lance snake at 1.2m–2m in length is an extremely venomous and is found in tropical America and tropical Asia.

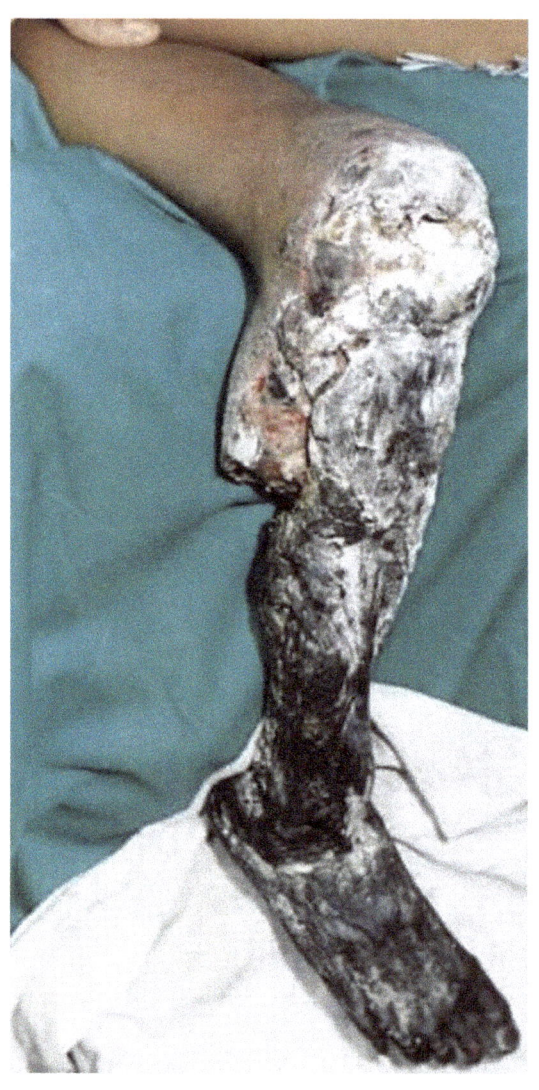

This boy has been bitten on the leg by the Fer de Lance snake. The venom contains a toxic chemical that causes necrosis – the destruction and death of the cells of the body. The leg would have to be amputated.

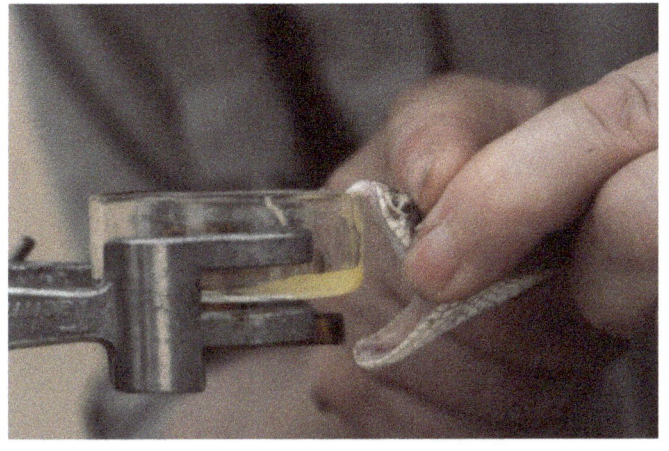

Milking snakes for their venom is an extremely dangerous and very skilful job. A degree in biology or biochemistry is required. A snake milker earns about $2,500 a month.

22

Constrictor Snakes

If you want to think of really big snakes, then it has to be the python and the anaconda. Strangely, it can be quite difficult to measure the length of a very long snake to try and find the world's longest. Snakes can be much longer if they live in zoos because they are well fed. Those in the wild must find food and may be diseased. The longest recorded length of a snake – the reticulated python, ever recorded was 'Medusa' that lived in a zoo in Kansas City, USA, and was 7.67m in length (25' 2"). There have been reports of an anaconda being seen in Brazil that was 10m (33') in length, but it was not scientifically measured. There is still a reward of $50,000 if a person can find a living snake that is longer than 30 feet (9.1m)!

There are two groups of the constrictor snakes. Firstly, the boas, which includes the anacondas and give birth to live young and live in Central and South America. Secondly the pythons which lay eggs and live in Asia Africa and Australia. Both the boas and the pythons are famous for the large prey they can capture and squeeze to death before eating them.

These snake use stealth and ambush to find their prey their prey and then bites and holds the prey using its many teeth.

The snake then coils its body around the prey and then uses its immense strength to squeeze its prey to death. Contrary to popular belief the snake does not crush the bones of its prey, but instead it restricts blood flow and prevents breathing in its prey. The snake can detect the heartbeat of its prey and when this stops the snake stops further squeezing.

The snakes have only one lung which is about a third of the length of the animal. Biologists have often thought that when the snake is killing its prey why do the constricting muscles of the snake not stop it from breathing? It is now known that the snake has "localised breathing" and because of the length of the lung only short lengths of the lung are squeezed when the prey is being captured, and the rest of the lung acts as a kind of reservoir for air.

Although it is very rare for the large constrictor snakes to kill and swallow humans there are records of such behaviour taking place. Almost every year there are reports of a person being eaten by a large snake, often a python. In October 2022 a fifty-year-old woman in Indonesia who was on her way to work went missing. Locals searched the area and found a 7m python suspected of eating the woman. When the snake was cut open the unfortunate woman was found dead inside the animal.

The world's longest snake in captivity – a Reticulated python called 'Medusa', measuring 7.67m (25' 5") in length and weighing 158kg (350lbs).

The skeleton of a Burmese python. The snake was found in the Everglades, Florida, USA. These pythons were illegally introduced into the Everglades because they were originally too big to be kept as pets! They reproduced very quickly and are now destroying much of the wildlife of the Everglades. It took this lady 250 hours of work to assemble 872 ribs and vertebrae, and 100 teeth of this 5.4m (17' 72") Burmese python.

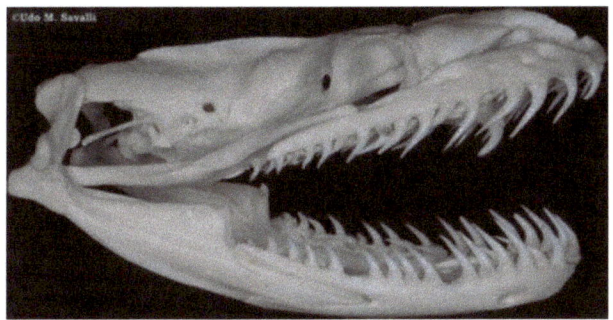

The skull of a python showing the large number of teeth which the animal uses to grab its prey.

In this zoo in Indonesia, it is possible for visitors to handle large pythons.

A captured python in Indonesia.

An Olive python eating a crocodile. These pythons are found in
Australia and reach 4m (13') in length, but occasionally reach
5m (17').

A captured python in Indonesia.

Anacondas are massive snakes. They are not the longest, but are the heaviest. Found in Brazil they can often reach 7m (23') in length. An anaconda discovered on a building site in Brazil was *estimated* to be 10m (33') in length 0.9m (3') wide and weighing 399kg (880lbs). It had to be moved using a crane!

23

Up the Nose and the Brain-Eating Amoeba

Living in the mud and silt at the bottom of very warm or hot freshwater ponds is a very nasty and horrible organism – the brain-eating amoeba. It is a parasitic single celled organism about three times smaller than the diameter of a human hair. It is a very strange organism because it can exist in three different forms. Sometimes the amoeba has two long hairlike structures called flagella. If a person enters water where the amoeba is living it is possible for water to enter the nose and for the amoeba to enter the nose using its flagella for locomotion. Once inside the nose the amoeba now changes into a different form – the feeding form. It takes between one and two hours to change from the flagella form to the feeding form. The feeding form now moves from the nose to the brain where it feeds on the brain cells of the infected person. Symptoms which appear in about five days of the infection include headaches, high temperature, drooping eyelids and changes in behaviour and mental state.

Although the infection is very rare it is nearly always fatal. From the years 1962 to 2022 only 157 infections were recorded, but only four people survived! A person does not become infected if the amoeba is swallowed and it cannot live in seawater. It is not contagious – it cannot be passed from one person to another. Sometimes warning signs are erected in areas where the Amoeba is known to exist.

The amoeba grows best in waters with a temperature of 46°C. With climate change and global warming occurring it is thought that there will be an increase in temperatures of cooler water and this will in turn increase the number of cases of the infection.

The Amoeba can exist in a third from known as the cyst form. In this form the Amoeba grows a tough protective coat around itself that protects the Amoeba from adverse conditions, e.g. low temperatures, chemical pollution or even a pond drying up. When the conditions improve the cyst opens and the Amoeba escapes from its protective coat.

The scientific name for the brain-eating amoeba is Naegleria fowleri and was named after an Australian doctor, Malcolm Fowler, who first found the parasite in children in 1965 at Adelaide Children's Hospital.

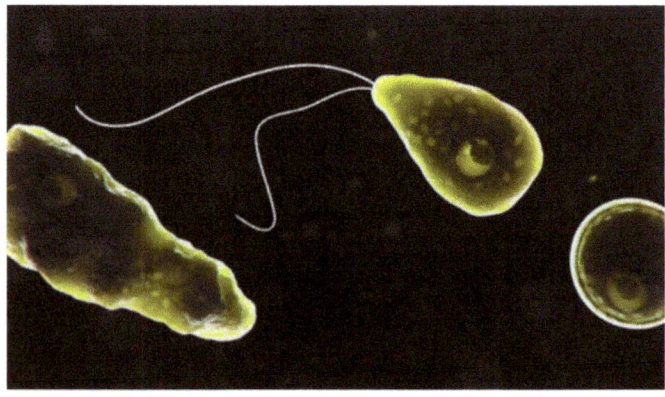

Photographs taken through the microscope of the three stages of the brain eating Amoeba. The Amoeba with the two long flagellae is the form that enters the nose. The stage that is long, including the oval shape with three appendages is the form that eats the brain cells. The spherical shape is the cyst stage that protects the Amoeba if the environmental conditions change.

The amoeba is common in the warmer parts of the USA and warning signs have been erected beside warmwater pools.

A photograph of the Roman Baths in the city of Bath, Somerset, England. In October 1978, these baths were closed because a child died from Naegleria fowleri after swimming in this Roman Bath. Even today the Roman Bath remains closed for swimming.

24

Army Ants –
Useful in an Emergency

There are over 200 species of army ants – some species live in nests, but some species do not and are always on the move. Army ants have a fearsome reputation and when on the move a column of these ants can be over 100m in length, 25m wide and contain over 50 million individuals. A column as large as this can eat 500,000 prey animals a day! Their food is mainly insects and spiders but are also known for feeding on small reptiles and birds.

Army ants, the largest of all the 12,000 species of ants, have very large mandibles (pincers). Imagine you are in the jungle, and you cut yourself very badly and there is no medical facilities available to stitch your cut. For over 3,000 years people living in the jungles of Africa have used army ants to close wounds and stop bleeding. The ant is held with each of its mandibles on either side of the wound. Because the ant feels threatened it closes its mandibles and this pinches the two ends of the cut together. The body of the ant is now cut-off leaving the head and mandibles attached to the skin. This procedure is repeated, depending upon the length of the cut. The mandibles remain closed for several days.

Most army ants are blind or have very poor eyesight. They do not have a permanent nest like other species of ant. In the colony of army ants different ants have different jobs. The worker ants forages for food; the large soldier ants have large mandibles and their job is to protect the colony from predators; male ants who fly from the colony to find a queen ant to mate with; finally the queen ant that is at the centre of the colony and controls everything that happens in the colony. If the queen ant dies the whole colony die with her!

The Army ant, sometimes called the safari ant, the driver ant or siafu (the east African Swahili word for ant)

The army ant has very long powerful mandibles which can be used for sutures (stitches) in an emergency.

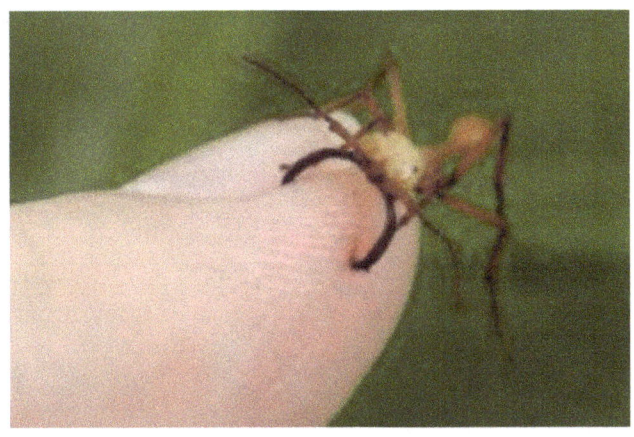

The mandibles of the army ant just beginning to pierce the skin.

Army ants on the move. A column of ants on the move may contain 50 million individual ants.

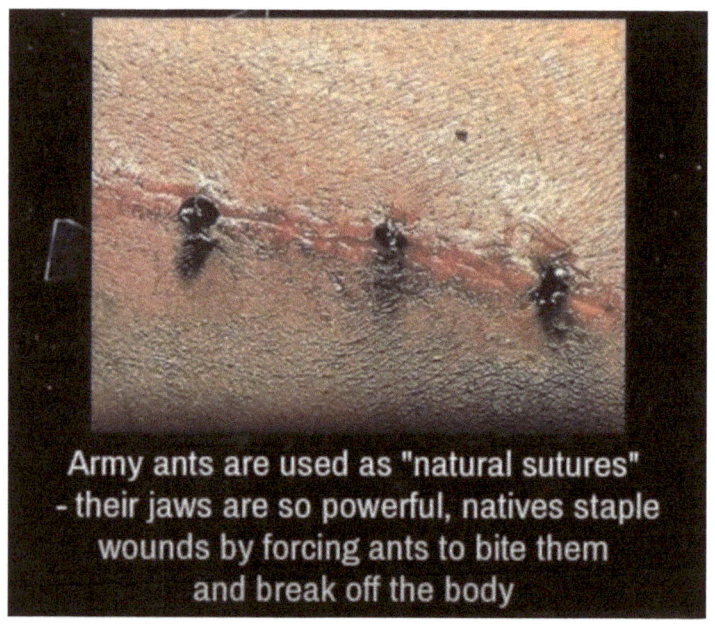

Army ants are used as "natural sutures"
- their jaws are so powerful, natives staple
wounds by forcing ants to bite them
and break off the body

A nasty cut in the skin has been stitched using the mandibles of the army ant. The black areas in the cut are the heads of the ants that have remained after the bodies of the ants have been cut off.

25

Conscientious Objectors, but Very Brave

How would you like to wear the clothes of a person who was suffering from a very nasty disease? How would you like to sleep in the bed of a person who had suffered from a very nasty disease? In 1940 this was exactly what people volunteered to do to be part of research into a disease called scabies. The scabies mite is a very small spider that burrows mainly underneath the skin of the hand and wrist but can also be found on the elbow and even the feet. If a person has the scabies mite, intense itching of the skin occurs around the area of the burrowing mite. Because the itching is so great the infected person scratches so severely that the skin breaks and bleeding occurs.

During World War 2, so many people in Britain had scabies that the Ministry of Health decided that research should be undertaken to find out why scabies was spreading throughout the population. In 1941 it was though that 2 million people in Britain suffered from scabies. There were ointments that could be rubbed onto the skin to kill the scabies mite, but unfortunately, they had a bad effect on the skin.

Children with scabies were excluded from school. Some children were so badly infected from scabies that they never attended school and were illiterate!! The research required volunteers who were prepared to wear the clothes and sleep on the mattresses and be covered in bed clothes in the beds of people who had scabies. At all times during the research discussions took place between the volunteers and the scientists so that the volunteers knew what was involved in the experiments. The volunteers came from a group of people called conscientious objectors. These were people who refused to fight for their country during the war. They often thought that war was wrong, or they had strong religious beliefs. They were often labelled as cowards. However, many conscientious objectors were very brave indeed – stretcher bearers in the field of battle, bomb disposal, fire fighters and ambulance drivers.

The research began on 1st December 1940 when beds, and clothing from infected people, laboratory equipment etc was installed in a large house in the city of Sheffield. Twelve volunteers were involved in the research. Each was paid about £120 and laboratory staff and technicians were also paid. Even a horse was kept at the cost of £20. After days and weeks of living in these disgusting, horrible conditions, ***not one volunteer became infected with scabies!!*** At the time scabies was thought to be spread by soldiers returning to Britain during the war and then passing on the mites to their families. People thought that they caught the disease by living close together in air-raid shelters. Others thought they caught it from lavatory seats, tables, and chairs and even from money, but the research showed that these were of no importance. They did find that scabies had a very long incubation time –

longer than 2 months. In other words, if you slept in the same bed as a person who had scabies, it would take at least 2 months for the person to develop the symptoms of the disease. The scabies mite tended to be killed at higher temperatures. They survived at lower temperatures found at the time in poor quality housing. The researchers found it difficult to collect live adult scabies mites for the experiments. They therefore used a mite that is found on horses which were more easily caught. In 2015, it was estimated that at any one time 204 million people in the world were infected with the scabies mite. Safe creams are now available to control and kill the scabies mite.

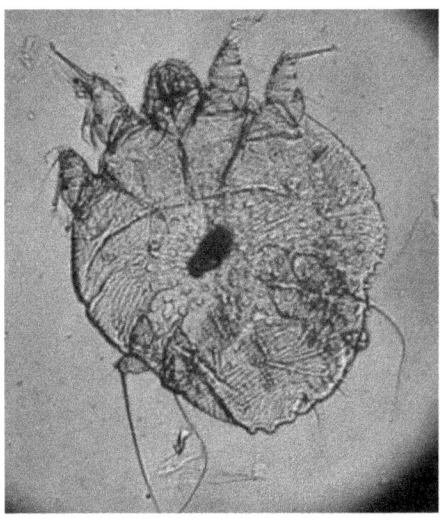

The scabies mite it is thought that at any one time about 204 million people are infected with this animal.

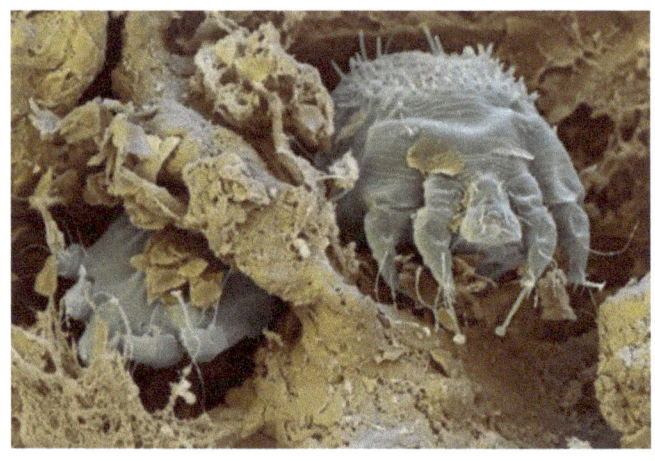

An electron microscope photograph of the scabies mite.

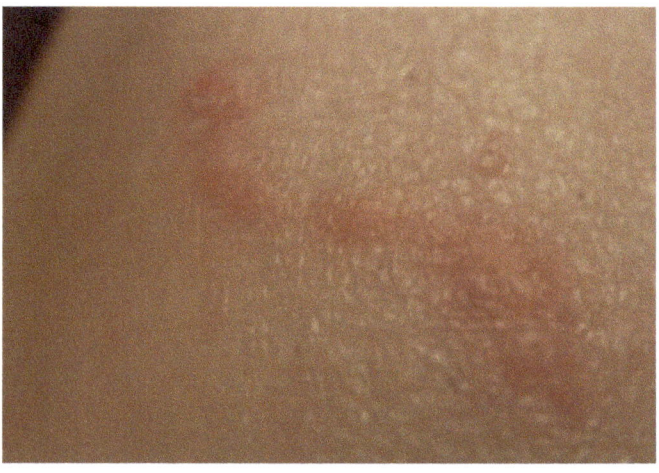

Three photographs showing scabies infection of the skin. Note the
long 'tunnels' made by the mite as it burrows under the skin.

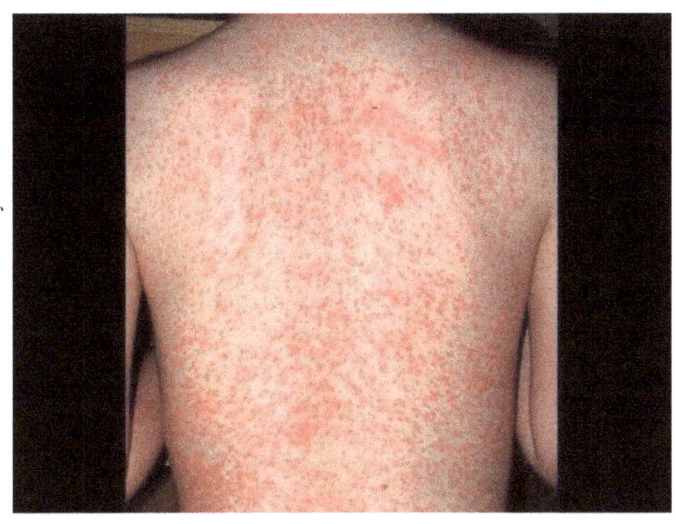
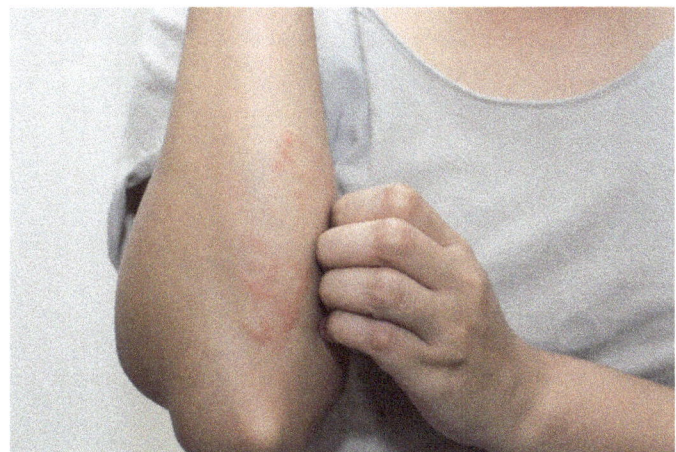

Websites

https://www.ncbi.nlm.nih.gov/pmc/articles/PMC4320518/figure/Fig1/

https://www.lommelegen.no/images/74449195.jpg?imageId =74449195&width =640&height=384&co

https://pixels.com/featured/2-demodex-folliculorum-eye-of-science.html https://premiumvisionsc.com//wp-content/uploads/2015/04/demodex5.jpg

https://www.bbc.co.uk/news/uk-england-37320399

https://www.bbvaopenmind.com/en/science/leading-figures/james-lind-and scurvy-the-first-cl

https://www.sciencehistory.org/files/dm-scurvy-prospectorsjpg

https://commons.wikimedia.org/wiki/File:Scurvy;_male_figure._Wellcome_M 0002829.jpg

https://www.primehealthchannel.com/wp-content/uploads/2013/04/Scurvy Image.b197b0.webp

http://www.epi.umn.edu/cvdepi/wp-content/uploads/2011/05/Hales-Horse.jpg

https://allthatsinteresting.com/botfly-larvae

https://allthatsinteresting.com/botfly-larvae

https://commons.wikimedia.org/wiki/File:D._hominis_adult

_female. png https://allthatsinteresting.com/botfly-larvae
https://bugsinourbackyard.org/the-human-botfly-dermatobia-hominis/ https://allthatsinteresting.com/botfly-larvae
https://www.geograph.org.uk/photo/2510649
https://upload.wikimedia.org/wikipedia/commons/2/27/Gian t_hogweed%2C_ Minnowburn%2C_Belfast_-_geograph.org.uk_-_3053893.jpg
https://www.zmescience.com/science/news-science/giant-hogweed-toxic-plant 3243/
https://wellcomecollection.org/works/x5b5q784/images?id=kev3tx3m www.rmg.co.uk/stories/blog/curatorial/removing-bladder-stone-size-tennis-ball 93
https://www.loc.gov/item/94505148/
https://commons.wikimedia.org/wiki/File:St_Martin_Alexis. jpg
https://commons.wikimedia.org/wiki/File:Paraponera_clavat a_(14500014836). jpg
https://www.bing.com/images/search?view=detailV2&ccid=q35CPHDI&id=55
0B8D7EC2AE87DA6BE53FAF394437C8268AB592&thid=OIP.q35CPHDIL
AjhYGJv8u3YRQHaIO&mediaurl=https%3a%2f%2fhips.he arstapps.com%2fe sq.h-cdn.co%2fassets%2f15%2f33%2f1439482799-justin-schmidt
1.jpg%3fresize%3d480%3a*&cdnurl=https%3a%2f%2fth.bi ng.com%2fth%2fi
d%2fR.ab7e423c70c82c08e160626ff2edd845%3frik%3dkr WKJsg3RDmvPw
%26pid%3dImgRaw%26r%3d0&exph=533&expw=480&q =justin+schmidt+a

nd+bullet+ant&simid=608021907788413059&FORM=IRP
RST&ck=4B0B26
EFCB3BCB2D12E25C548B2511E6&selectedIndex=0&idp
p=overlayview&aj axhist=0&ajaxserp=0
https://commons.wikimedia.org/w/index.php?search=bullet+
ant+and+stinger&t
itle=Special:MediaSearch&go=Go&type=image
https://www.troab.com/amazon-tribes-painful-rituals-to-
prove-adulthood/
https://commons.wikimedia.org/w/index.php?fulltext=Searc
h&search=File%3
BScorpion+Photograph+By+Shantanu+Kuvesk&title=Speci
al:Search&ns0=1& ns6=1&n
https://commons.wikimedia.org/wiki/File:Emperor_Scorpio
n_in_hand_John. jpg
https://commons.wikimedia.org/wiki/File:Parabuthus_transv
aalicus_(male).jpg https://factzoo.com/book/deathstalker-
yellow-poison-stinger-desert/
https://commons.wikimedia.org/wiki/File:Avispa_marina_cr
opped.png
https://commons.wikimedia.org/wiki/File:Box_jellies_over_
sand_at_Geldkis_ DSC00331.JPG
https://wetsuitwarehouse.com.au/collections/stinger-
suits/products/adrenalin adults-hooded-lycra-stinger-suit
https://en.wikipedia.org/wiki/File:Marinesting1.jpg
https://boxjellyfish.org/australian-box-jellyfish-facts-
natures-deadliest-creature/ http://imgkid.com/box-jellyfish-
sting-scars.shtml
https://commons.wikimedia.org/wiki/File:Hapalochlaena_fa
sciata_Toba1. jpg
https://commons.wikimedia.org/wiki/File:Blue_Ringed_Oct

opus.png

https://commons.wikimedia.org/wiki/File:Hapalochlaena_lu nulata2.JPG 94 https://www.animalspot.net/wp-content/uploads/2015/05/Blue-Ringed Octopus-Size.jpg

https://commons.wikimedia.org/wiki/File:Goldenergiftfrosc h1cele4.jpg

https://commons.wikimedia.org/wiki/File:Yellow banded.poison.dart.frog.arp.jpg

https://commons.wikimedia.org/wiki/File:Ameerega_trivittat a_(Madre_de_Dio s,_Peru).jpg

https://commons.wikimedia.org/wiki/File:Strawberry_poiso n dart_frog_(Oophaga_pumilio_or_Dendrobates_pumilio)_(94 29685305).jpg

https://commons.wikimedia.org/wiki/File:Red_poison_dart_ frog_(cropped).jpg

https://commons.wikimedia.org/wiki/File:Erythrolamprus_e pinephalus_327316 32.jpg

https://commons.wikimedia.org/wiki/File:Dendrocnide_mor oides_foliage_SF2 0326.jg

https://www.snexplores.org/article/australian-stinging-tree-touch-pain-toxin gympie-gymp

https://www.snexplores.org/article/australian-stinging-tree-touch-pain-toxin gympie-gymp

https://allthatsinteresting.com/gympie-gympie

https://www.science.org/doi/10.1126/sciadv.abb8828

https://commons.wikimedia.org/wiki/File:Red-legged_Pademelon.jpg https://www.abc.net.au/news/2021-05-25/mawson-ill-fated-far-eastern-partys antarctic-voyage-memorial/100161834

https://collection.sl.nsw.gov.au/record/Yr86777n – viewer

135

https://factsforantarctica.weebly.com/map-of-mawsons-journey-in antarctica.html

https://commons.wikimedia.org/wiki/File:Douglas_Mawson_recuperating. jpg

https://commons.wikimedia.org/wiki/File:Atropa_bella-donna_sl5.jpg

https://commons.wikimedia.org/wiki/File:20190830Atropa_belladonna1. jpg

https://commons.wikimedia.org/wiki/File:Eye_treated_with_dilating_eye_drop s.jpg

https://commons.wikimedia.org/wiki/File:Nepenthes_rajah.png

https://www.bing.com/images/search?view=detailV2&ccid=yBFsOesX&id=18

2FEE27733FEA02842609DA22CA9E41ABA9EE56&thid=OIP.yBFsOesXfRI

NpidzYdUcyQHaLH&mediaurl=https%3a%2f%2fallthatsinteresting.com%2f 95 wordpress%2fwp-content%2fuploads%2f2018%2f04%2fpitcher-plant lizard.jpg&cdnurl=https%3a%2f%2fth.bing.com%2fth%2fid% https://www.ripleys.com/weird-news/shrew-poo-plant/?ref=prevpost

https://commons.wikimedia.org/wiki/File:Hyla_cinerea_pitcher_plant.jpg

https://commons.wikimedia.org/wiki/File:Dionaea_muscipula_Royal_Red_Ven us_Fly_Trap.jpg

https://commons.wikimedia.org/wiki/File:Venus_Flytrap_showing_trigger_hair s.jpg

https://commons.wikimedia.org/wiki/File:Dionea_in_action. jpg

https://commons.wikimedia.org/wiki/File:Dionaea,_muscoid

_fly.jpg
https://commons.wikimedia.org/wiki/File:Bladderwort_(Utri
cularia)_in_bloom,
_Cass_County,_Texas,_USA_(April_2017).jpg
https://commons.wikimedia.org/wiki/File:Common_Bladder
wort_(3629471561).jpg
https://commons.wikimedia.org/wiki/File:Utricularia_vulgar
is_001.JPG
https://commons.wikimedia.org/wiki/File:Utricularia_aurea_
8_Darwiniana.jpg https://carnivorousplantresource.com/the-
plants/bladderwort/
https://commons.wikimedia.org/wiki/File:Drosera_rotundifo
lia_Rosiczka_okr %C4%85g%C5%82olistna_2022-06-
21_06.jpg
https://commons.wikimedia.org/wiki/File:Drosera_anglica_n
e2.jpg https://commons.wikimedia.org/wiki/File:Round
leaved_Sundew_(Drosera_rotundifolia)_with_ant_prey_(94
10495122).jpg
https://commons.wikimedia.org/wiki/File:Drosera_capensis_
bend.JPG
https://commons.wikimedia.org/wiki/File:Pulex_irritans_ZS
M.jpg
https://commons.wikimedia.org/wiki/File:Female_human_h
ead_louse.jpg
https://commons.wikimedia.org/wiki/File:Ixodes_scapularis.
jpg /media/File:Ixodes_scapularis.jpg
https://www.realclearscience.com/lists/gutwrenching_pictur
es_of_parasitic_dis eases/pinworm.html - !
https://commons.wikimedia.org/wiki/File:Sarcoptes_scabei_
2.jpg NOT Available
https://commons.wikimedia.org/wiki/File:Fleabite.JPG

https://www.homenaturalcures.com/head-lice-symptoms-causes/

https://commons.wikimedia.org/w/index.php?search=lyme+disease+bite&title=

Special:MediaSearch&go=Go&type=image

https://pharmamum.com/worms-in-children/ 96

https://commons.wikimedia.org/wiki/Category:Scabies -/media/File:ScabiesD08.JPG

https://commons.wikimedia.org/wiki/File:2015.11.26.192558_Cimex_lectulari us_bites.jpg -/media/File:2015.11.26.192558_Cimex_lectularius_bites.jpg

https://www.dailymail.co.uk/news/article-9397203/Thai-man-excretes-59 FOOT-tapeworm-doctors-visit-extreme-flatulence.html https://eyewiki.org/File:Loa_loa_external.jpg

https://commons.wikimedia.org/wiki/File:Elephantiasis_of_t he_leg_(ATED_74-6426-

2),_National_Museum_of_Health_and_Medicine_(2837011 06).jpg

https://commons.wikimedia.org/wiki/File:Aedes_aegypti_C DC8936.tif

http://www.forestryimages.org/browse/detail.cfm?imgnum= 5390481

https://plantdoctor.co.nz/assets/uploads/2016/05/nematode-fungal-hyphae.jpg

https://commons.wikimedia.org/wiki/File:Cordyceps_matou _lagarta.jpg

https://commons.wikimedia.org/wiki/File:Grasshopper_cord yceps_(198852840 90)

https://commons.wikimedia.org/wiki/File:Ant_Killed_by_Fu ngus__Cockscomb_Wildlife_Sanctuary,_Belize.jpg

https://commons.wikimedia.org/w/index.php?search=mount

ain+chicken+frog &title=Special:MediaSearch&go=Go
https://commons.wikimedia.org/wiki/File:Chytridiomycosis.
jpg https://www.livescience.com/19628-fungal-diseases-
emerging-threat.html.